Soil Nematodes of Grasslands in Northern China

Academic Press is an imprint of Elsevier
125 London Wall, London EC2Y 5AS, United Kingdom
525 B Street, Suite 1800, San Diego, CA 92101-4495, United States
50 Hampshire Street, 5th Floor, Cambridge, MA 02139, United States
The Boulevard, Langford Lane, Kidlington, Oxford OX5 1GB, United Kingdom

Notices
Knowledge and best practice in this field are constantly changing. As new research and experience broaden our understanding, changes in research methods, professional practices, or medical treatment may become necessary.

Practitioners and researchers may always rely on their own experience and knowledge in evaluating and using any information, methods, compounds, or experiments described herein. In using such information or methods they should be mindful of their own safety and the safety of others, including parties for whom they have a professional responsibility.

To the fullest extent of the law, neither the Publisher nor the authors, contributors, or editors, assume any liability for any injury and/or damage to persons or property as a matter of products liability, negligence or otherwise, or from any use or operation of any methods, products, instructions, or ideas contained in the material herein.

Library of Congress Cataloging-in-Publication Data
A catalog record for this book is available from the Library of Congress

British Library Cataloguing-in-Publication Data
A catalogue record for this book is available from the British Library

ISBN: 978-0-12-813274-6

For information on all Academic Press publications visit our website at
https://www.elsevier.com/books-and-journals

  Working together
to grow libraries in
developing countries

www.elsevier.com • www.bookaid.org

Publisher: Sara Tenney
Acquisition Editor: Kristi Gomez and Xiufang WU
Editorial Project Manager: Pat Gonzalez
Production Project Manager: Julia Haynes
Designer: Vicky Pearson

Typeset by TNQ Books and Journals

Contents

List of Contributors   vii
Foreword   ix
Preface   xi

1. **Introduction**
    1.1 **Grasslands in China**   1
    1.2 **Background Information of the Field Survey Across the
         Grassland Transect**   4
         References   10

2. **Nematode Diversity and Distribution Along the
    Grassland Transect**
    2.1 **Nematode Community Composition Along the Grassland
         Transect**   13
    2.2 **Patterns of Climatic Distribution of Nematode Trophic
         Groups**   18
    2.3 **Nematode Generic Richness and Diversity**   29
    2.4 **Relations of Nematode Assemblage Composition With
         Plant and Soil Properties**   39
         References   40

3. **Nematode Genera and Species Description Along
    the Transect**
    3.1 **Background on Nematode Taxonomy**   45
         3.1.1 General Information on Soil Nematodes' Life Strategies   45
         3.1.2 General Morphological Terminologies and Structures of
                Nematodes   47
    3.2 **General Characteristics of Nematode Families Recorded
         From the Grassland Transect**   63
    3.3 **Nematode Genera and Species Present in the Grassland
         Transect**   68
         References   224

4.   Advances and Perspectives in Soil Nematode Ecology in China

4.1	**Progress and Research Field in Soil Nematode Ecology**	**229**
4.2	**Future Perspectives of Soil Nematode Ecology in China**	**233**
	4.2.1 Connection of Soil Nematodes With Other Soil Biota Within Food Webs	233
	4.2.2 Feedback Between Aboveground and Belowground Communities	233
	4.2.3 Molecular Analysis of Soil Nematode Diversity	234
	References	234

Index   **239**

List of Contributors

Qi Li, Institute of Applied Ecology, Chinese Academy of Sciences, Shenyang 110016, China

Wenju Liang, Institute of Applied Ecology, Chinese Academy of Sciences, Shenyang 110016, China

Xiaoke Zhang, Institute of Applied Ecology, Chinese Academy of Sciences, Shenyang 110016, China

Mohammad Mahamood, Institute of Applied Ecology, Chinese Academy of Sciences, Shenyang 110016, China

Jun Yu, Institute of Applied Ecology, Chinese Academy of Sciences, Shenyang 110016, China

Siwei Jiang, Institute of Applied Ecology, Chinese Academy of Sciences, Shenyang 110016, China

XiaoTao Lü, Institute of Applied Ecology, Chinese Academy of Sciences, Shenyang 110016, China

Xuelian Bao, Institute of Applied Ecology, Chinese Academy of Sciences, Shenyang 110016, China

Xiaoming Sun, Institute of Applied Ecology, Chinese Academy of Sciences, Shenyang 110016, China

Ying Lü, Institute of Applied Ecology, Chinese Academy of Sciences, Shenyang 110016, China

Dan Xiong, Institute of Applied Ecology, Chinese Academy of Sciences, Shenyang 110016, China

Cunzheng Wei, Institute of Botany, Chinese Academy of Sciences, Beijing 100093, China

Xingguo Han, Institute of Applied Ecology, Chinese Academy of Sciences, Shenyang 110016, China

Foreword

Belowground biodiversity represents one of the largest reservoirs of biological diversity on Earth and has a key role in determining the ecological and evolutionary response of terrestrial ecosystems to current and future environmental change. Grassland contains an abundant and diverse fauna belowground, and nematodes constitute a major group that performs important functions such as feeding and dispersing both saprophytic beneficial pathogenic bacteria and fungi, as well as regulating the amount of inorganic nitrogen available to plants, in addition to feeding directly on plant roots.

Although much insight has been provided into aboveground ecosystems, the belowground biotic community and its role in grassland ecosystems have received relatively little attention. The current volume by scientists from the Institute of Applied Ecology, Chinese Academy of Sciences, is a courageous effort in this direction. Profs. Qi Li, Wenju Liang, Xiaoke Zhang, and Dr. M.D. Mahamood have provided a detailed account of the distribution pattern of soil-inhabiting nematodes along a 3600-km-long grassland transect of northern China and have tried to evaluate the relationship of nematodes to climatic conditions, plant biomass, and soil characteristics. The main strength of the book is in its detailed taxonomic information on the large number of nematode genera recorded representing various trophic groups. No other book from China has provided such detailed information on grassland nematode fauna.

In recent years, most soil ecologists have strongly emphasized soil nematodes in their ecological studies. The current volume will be highly beneficial to this scientific community. I must congratulate Prof. Li and his coworkers for their effort in bringing out this publication, which will be of great importance to soil ecologists and nematologists for years to come.

Wasim Ahmad

Preface

The term "nematode" is derived from two Greek words: *nema* (thread) and *eidos* (like). To many of us, nematodes are something unheard or unseen because they are tiny organisms that cannot be seen with the naked eye. However small nematodes are, they occupy an important position in the food chain of the soil ecosystem. Compared with aboveground ecosystems, relatively little attention has been paid to the belowground biotic communities of grassland ecosystems in China, and the limited research that exists has been conducted mainly on a relatively small scale. Until now, our knowledge about nematode diversity in larger-scale patterns has still been scanty, especially at the genus or species level. Because of variations in life strategies and feeding preferences within each family, research at the genus or species level is important to a better understanding of the soil biodiversity and function they provide.

In this book we present the diversity and distribution patterns of nematodes along the 3600-km grassland transect of northern China and evaluate their relations to climatic conditions, plant biomass, and soil characteristics for a better understanding of soil nematode distribution, and the consequences they may exhibit for ecological functioning. In addition, detailed taxonomic information is presented on 66 nematode genera belonging to 32 families along the grassland transect. Finally, we briefly review advances in and perspectives on research in soil nematode ecology in China. We hope this preliminary research will facilitate better understanding of the diversity and distribution of soil biotic communities in grassland ecosystems of China, and that it will help determine potential factors that drive soil biodiversity distribution and sustainable management of grassland ecosystems under future climate change conditions.

We thank Prof. Xingguo Han and all of the members of the Shenyang sampling campaign teams including Dr. Xiaobo Wang, Chao Wang, Dongwei Liu, Wentao Luo, Jiao Feng and Xiaoguang Wang from the Institute of Applied Ecology, Chinese Academy of Sciences, and Dr. Quansheng Chen from the Institute of Botany for their assistance with field sampling. We gratefully acknowledge Prof. Wasim Ahmad for his comments on the earlier version of this book. This research was supported by the National Natural Science Foundation of China (31570519) and the Strategic Priority Research Program of the Chinese Academy of Sciences (XDB15010402).

Wenju Liang

Chapter 1

Introduction

1.1 GRASSLANDS IN CHINA

In a narrow sense, "grassland" is defined as "ground covered by vegetation dominated by grasses, with less than 10 percent or no tree/shrub cover" (Suttie et al., 2005). It is naturally distributed in areas with an annual amount of precipitation between that of a desert and forest on all continents except Antarctica (Allaby, 2012; Woodward et al., 2004). As one of the primary terrestrial ecosystems, grasslands occupy about a quarter of the world's land surface (Harvey, 2001) and provide many important ecological services, many of which are mediated by the soil biota, such as soil structure maintenance, water regulation, nutrient cycling (Bardgett and Chan, 1999), and plant production (Bardgett and Wardle, 2010). There are various types of grassland, according to their regional environment and vegetation composition, including the Eurasian Steppe, the North American Prairie, the South American Pampas, and the African and Australian Savanna (Wu et al., 2015).

Grasslands in China constitute an integral part of the Eurasian Steppe, the world's largest grassland (Kang et al., 2007). The total grassland area of China is reported to range from 2.20 million km^2 (the Bulletin of Land and Resources in China, 2014) to 4.06 million km^2 (Hou, 1982), covering 22.9%–42.3% of China's land (Fang et al., 2010; Kang et al., 2007). Such a huge variation in the estimated grassland area is probably caused by different definitions of "grassland" and by varying data sources and calculation models adopted by different researchers. However, 3.93 million km^2 of total grassland area including 3.31 million km^2 of usable grassland is currently widely accepted (Fang et al., 2010; Hu and Zhang, 2006). Based on this estimation, after Australia and Russia, China has the third largest grassland area in the world (Hu and Zhang, 2006).

From a geographical perspective, China's grassland occurs mainly in the northern temperate regions and the alpine areas on the Tibetan Plateau (Fan et al., 2008). Grassland in these regions accounts for approximately 70% of the country's total grassland area. The provinces and autonomous regions with a grassland area, which range from 15.3 to 82.1 million hectares, are located mainly in Tibet, the Inner Mongolia Autonomous Region (Inner Mongolia), the Xinjiang Uyghur Autonomous Region (Xinjiang), Qinghai, Sichuan, and Gansu and Yunnan provinces (Hu and Zhang, 2006). According to the "vegetation-habitat

Soil Nematodes of Grasslands in Northern China. http://dx.doi.org/10.1016/B978-0-12-813274-6.00001-6

classification system," grasslands in China can be divided into nine classes (Hu and Zhang, 2006). Among them, the temperate steppe, alpine meadow, alpine steppe, temperate desert, and temperate meadow are five dominant classes that in total account for 74.8% of China's grassland (Table 1.1).

In China, the history of grassland survey and research can be summarized into four periods: pre-1950s, 1950–75, 1976–95, and 1996 to the present (Kang et al., 2007); most of the key research initiatives on the grassland ecosystem took place after the mid-1990s (Kang et al., 2007). The latest research efforts mainly focused on grassland management for sustainable productivity and ecosystem stability, and changes in structure and function in the context of global change, such as community structure and succession (Bai et al., 2004), biogeochemistry (Yang et al., 2014; Yu et al., 2010), carbon storage/sequestration (Fang et al., 2010), greenhouse gas dynamics (Chen et al., 2003), and pest (e.g., grasshopper) and rodent control and management (Kang et al., 2007).

This research provides much insight into aboveground vegetation; there is continuing effort to understand belowground processes (Kang et al., 2007). Until now, however, relatively little attention has been paid to belowground biotic communities of grassland ecosystems, most of which are on a relatively small scale (Chen et al., 2015). Although some patterns of spatial variation in biodiversity have been reported aboveground, we are far from achieving a clear picture of the occurrence of soil biota in terrestrial ecosystems (Decaëns, 2010), and knowledge of the distribution and abundance of some known species is still poor owing to the taxonomic deficit regarding soil organisms (Wilson and Knopt, 2002). So far, only a few studies have reported global patterns of soil biota distribution (Bardgett et al., 2005; Nielsen et al., 2014; Wardle, 2002). However, the potential mechanisms that control a given group of soil organisms may vary with different environmental conditions and different scales (Bardgett et al., 2005). Some fundamental questions remain to be resolved: What are the dominant drivers determining the distribution of soil biota at different scales? How does soil biodiversity respond to certain environmental gradients: for example, precipitation, temperature, and soil pH? Do dominant plant species or the composition of vegetation contribute to variations in the diversity and distribution of belowground biotic communities among different grassland types?

It is well-known that the belowground biota is important to maintain ecosystem functioning, such as nutrient and carbon cycling (Coleman et al., 2004) and primary production (Bardgett and van der Putten, 2014). Soil biota is closely linked with the aboveground subsystems. Feedback between the aboveground and belowground subsystems has a key role in regulating ecosystem processes such as plant community composition, vegetation succession, and net primary productivity of grasslands (Bezemer et al., 2006, 2010; van der Putten et al., 2013; Wardle et al., 2004). Therefore, knowing the diversity and distribution of soil biotic communities in different types of grassland in northern China will enable us to understand better how environmental factors and plant species

TABLE 1.1 Classes and Subclasses of Grassland and Corresponding Areas in China

Grassland Class	Subclass	Area (Hectares)	Relative to the Total Grassland Area (%)
Alpine meadow	Sum	63,720,549	16.22
Alpine steppe	Meadow-steppe	6,865,734	1.75
	Typical steppe	41,623,171	10.59
	Desert-steppe	9,566,006	2.43
	Sum	58,054,911	14.77
Alpine desert	Sum	7,527,763	1.92
Temperate steppe	Meadow-steppe	14,519,331	3.70
	Typical steppe	41,096,571	10.46
	Desert-steppe	18,921,607	4.82
	Sum	74,537,509	18.98
Temperate desert	Typical desert	45,060,811	11.47
	Steppe-desert	10,673,418	2.72
	Sum	55,734,229	14.19
Temperate meadow	Lowland meadow	25,219,621	6.42
	Mountain meadow	16,718,926	4.26
	Sum	41,938,547	10.68
Warm shrubby tussock	Tussock	6,657,148	1.69
	Typical tussock	11,615,910	2.96
	Sum	18,273,058	4.65
Tropical shrubby tussock	Tussock	14,237,196	3.62
	Typical shrub-tussock	17,551,276	4.47
	Savanna	863, 144	0.22
	Sum	32,651,616	8.31
Marsh	Sum	2,873,812	0.73

Adapted from Hu, Z.Z., Zhang, D.G., 2006. The Country Pasture/Forage Resources Profiles for China. http://www.fao.org/ag/AGP/AGPC/doc/Counprof/china/china1.htm.

influence the soil biota and, in turn, how their responses alter the ecosystem function and services they provide. Furthermore, this research on the distribution of soil biota on a larger scale will enable a clear understanding of the factors that drive the distribution of soil communities and provide useful information for local people to improve the productivity and sustainability of grasslands.

1.2 BACKGROUND INFORMATION OF THE FIELD SURVEY ACROSS THE GRASSLAND TRANSECT

To address the limitations of previous research on China's temperate grassland and to narrow the gap in knowledge, in 2012 the Institute of Applied Ecology, Chinese Academy of Sciences, conducted a field survey along a transect across the northern temperate grassland extending from the central-eastern Xinjiang Autonomous Region to the eastern part of Inner Mongolia in northern China (Figs. 1.1 and 1.2). Samples of grasses, shrubs, and soils were collected to measure net primary productivity, nutrient elements in soils and plants, and the community composition of soil microbes and nematodes. Because the major discoveries regarding soil microbial diversity and eco-function from this field survey have been published elsewhere (Feng et al., 2016; Luo et al., 2013, 2015; Wang et al., 2014, 2016, 2017), here we summarize the changes in soil nematode community composition and diversity across the grassland transect.

This transect is located near the Northeast China Transect (a constituent of the International Geosphere–Biosphere Program) and extends into Gansu and Xinjiang, covering about 3600 km in length (Fig. 1.1). The longitudes of

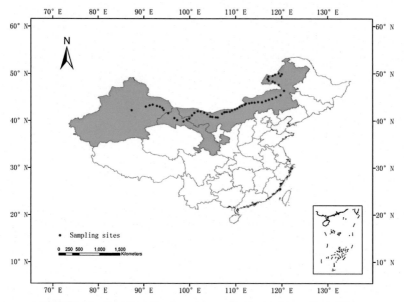

FIGURE 1.1 Sampling sites along the grassland transect in northern China.

this transect ranged from 87° 37′ E to 120° 48′ E and the latitude ranged from 40° 26′ N to 50° 01′ N. Along the transect, mean annual precipitation increased from 34 mm in Xinjiang (west) to 436 mm in Inner Mongolia (east), whereas the mean annual temperature (MAT) ranged from −2.9°C to 9.4°C. The four main vegetation types located in this transect were desert, desert steppe, typical steppe, and meadow steppe and primary productivity ranged from <10 g m^{-2} year^{-1} in the desert to >400 g m^{-2} year^{-1} in the meadow steppe (Wang et al., 2016). The richness of species per square meter ranged from 0 (no plants) in the west of the desert to >30 in the eastern parts of the meadow steppe (Lü et al., unpublished data). Related soil types of this region are gray-brown desert soil, brown calcic soil, and chestnut soil distributed from west to east. More details are given by Luo et al. (2013) and Wang et al. (2014). Precipitation is a crucial environmental factor determining the generation and distribution of grassland at the regional to continental scales. In parallel with an increase in precipitation, desert, desert steppe, typical steppe, and meadow steppe (the four main grassland types in temperate northern China) are distributed successively from northwest to northeast along this transect (Fan et al., 2008).

Desert ecosystems are mainly distributed in the Tarim Basin in south Xinjiang, northwest Gansu, and the west Inner Mongolian Plateau (Fig. 1.3). These areas are characterized by an arid continental climate with a mean annual precipitation less than 100 mm. The sparse vegetation mainly includes *Sympegma regelii, Salsola passerina, Reaumuria soongorica, Ceratoides latens, Kalidium schrenkianum, Potaninia mongolica, Nitraria sphaerocarpa, Ephedra przewalskii, Haloxylon persicum, Iljnia regelii, Nitraria sibirica, Halocnemum strobilaceum, Calligonum* spp., *Caragana* spp., and *Zygophyllum* spp. (Hu and Zhang, 2006; Ni, 2002).

Desert steppes are arid grasslands mainly distributed in the west of the Ordos Plateau, west of the Loess Plateau, and the Junggar Basin in north Xinjiang with

FIGURE 1.2   Fieldwork team for the grassland transect. *Photographed by Zhao Ying.*

FIGURE 1.3 Desert ecosystem along the transect. *Photographed by Lü X.T.*

FIGURE 1.4 Desert steppe along the transect. *Photographed by Lü X.T.*

mean annual precipitation between 100 and 250 mm (Hu et al., 2007; Ni, 2002; Sun, 2005) (Figs. 1.4 and 1.5). In these areas the soil type is brown-gray desert soil (Hu et al., 2007) and the dominant plant species are *Stipa gobica, Stipa klemenzii, Stipa breviflora, Stipa glareosa, Cleistogenes songorica, Agropyron desertorum, Allium polyrhizum, Hippolytia trifida, Ajania fruticulosa, Artemisia frigida, Salsola collina, Salsola laricifolia, Allium mongolicum, Nitraria tangutorum, Caragana* spp., *Ephedra* spp., *Reaumuria* spp., and *Suaeda* spp. (Hu et al., 2007; Luo et al., 2015; Ni, 2002).

Typical steppes are mainly located on the Inner Mongolian Plateau and the Loess Plateau, including midwest of the Xilin Gol Plateau, most of the Ordos Plateau, and west of the Loess Plateau (Hu et al., 2007; Kang et al., 2007; Sun, 2005) (Figs. 1.6 and 1.7). These areas are characterized by a semiarid climate

FIGURE 1.5 Sampling in a desert steppe along the transect. *Photographed by Lü X.T.*

FIGURE 1.6 Typical steppe along the transect. *Photographed by Lü X.T.*

FIGURE 1.7 Sampling in a typical steppe. *Photographed by Lü X.T.*

FIGURE 1.8 Sampling in a meadow steppe. *Photographed by Lü X.T.*

with hot wet summers and cold dry winters. The mean annual precipitation ranges from 250 to 400 mm and the MAT varies between −2°C and 2°C (Ni, 2002; Sun, 2005). The soil is generally classified as brown calcic and chestnut soil (Calcic Kastanozem, FAO). Plant species in this type of grassland exhibit drought tolerance and are dominated by *Stipa grandis, Stipa krylovii, Stipa bungeana, Stipa capillata, Festuca sulcata, Cleistogenes squarrosa, Agropyron cristatum, Leymus chinensis, A. frigida, Artemisia intramongolica, Artemisia gmelinii, Caragana microphylla, Achnatherum sibiricum, Allium tenuissimum, Koeleria cristata, Potentilla acaulis, Thymus mongolicus,* and *Ephedra* spp. (Hu et al., 2007; Luo et al., 2015; Ni, 2002; Wu et al., 2015).

The meadow steppe occurs naturally in moist and fertile regions (with high organic carbon content) in northeast China with mean annual precipitation ranging from 400 to 600 mm (Kang et al., 2007; Sun, 2005). Specifically, it is mainly distributed in the east of the Xilin Gol plateau, the east Hulunbuir plateau, and the southeast Songnen Plain (Hu et al., 2007) (Fig. 1.8). The MAT is between −3.0°C and 1.0°C (Ni, 2002). The most distributed soil types are chernozem and dark chestnut soil (Calcic Kastanozem, FAO) according to the Soil Classification System of China (Hu et al., 2007). The dominant plant species are *Stipa baicalensis, Bothriochloa ischaemum, Cleistogenes mucronata, L. chinensis, Leymus angustum, Filifolium sibiricum, Stipa grandis, Artemisia* spp., *Carex* spp., *Festuca* spp., and *Lespedeza* spp. (Hu et al., 2007; Kang et al., 2007; Luo et al., 2015; Ni, 2002).

Along the grassland transect, 66 sites were set up at intervals of 50–100 km with minimal grazing and other obvious anthropogenic disturbances. At each sampling site, two 50×50-m main plots representing local main landscapes were selected, with a distance of around 1–2 km between the two plots in each site and five 1 × 1-m subplots within each plot designated within the large plot. In each subplot, after removing aboveground plant tissue and litter, five soil

TABLE 1.2 Background Information and Soil Chemical Properties of the Four Types of Grasslands in Northern China

	Desert	Desert Steppe	Typical Steppe	Meadow Steppe
Sampling no.	5–26	27–44	45–49, 52–60, 62	50, 51, 61
Soil type	Aeolian sandy soil	Brown desert soil, gray desert soil	Brown calcic soil, chestnut soil	Chernozem, chestnut soil
Mean annual temperature (°C)	5.28–9.93	1.13–6.93	−1.85 to 1.32	−2.88 to 0.42
Mean annual precipitation (mm)	34–101	100–246	260–380	406–435
Elevation (m)	775–1731	956–1617	530–1104	754–972
Soil pH	7.69–9.16	7.67–8.76	6.52–7.76	6.74–6.81
Aridity	0.91–0.97	0.71–0.90	0.52–0.68	0.43–0.47
Total organic carbon (%)	0.06–0.60	0.14–0.61	0.92–4.35	1.91–5.02
Total nitrogen (%)	0.01–0.06	0.02–0.08	0.07–0.38	0.16–0.37
Total phosphorus (%)	0.02–0.08	0.01–0.03	0.02–0.07	0.02–0.80

samples were randomly collected at 0–10 cm using a soil corer (5 cm in diameter) and thoroughly mixed and pooled as a composite sample. Soil samples were passed through a 2.0-mm sieve and were then stored in a plastic bag at 4°C for nematode extraction and soil chemical property analyses. The geographical coordinates and altitude of each sampling site were recorded by a global positioning satellite device (eTrex Venture, Garmin, USA) (Wang et al., 2014). Grassland type, soil classification, and dominant plant species were recorded as well. Table 1.2 lists background information and soil chemical properties of the four types of grassland in northern China.

Previous research on soil biodiversity pointed out the dearth of information on the diversity of soil biota, especially at the genus or species levels (Bardgett and van der Putten, 2014; Decaëns 2010). Thus, in this book we first examine the nematode distribution patterns along the grassland transect at the trophic group, family, and genus levels and then evaluate their relationship to climatic conditions, plant biomass, and soil parameters. In Chapter 3, we present a detailed taxonomy of nematodes that exist along the 3600-km transect, for a better understanding of nematode diversity along the grassland transect in northern China. Finally, we review the advances and perspectives in soil nematode ecology research in China. We hope this preliminary research will help to gain insight in predicting the impact of global change on belowground soil biota, and will allow a better understanding of the functioning and services they provide in terrestrial ecosystems.

REFERENCES

Allaby, M., 2012. Oxford Dictionary of Plant Sciences, third ed. Oxford University Press, Oxford, UK.

Bai, Y.F., Han, X.G., Wu, J.G., Chen, Z.Z., Li, L.H., 2004. Ecosystem stability and compensatory effects in the Inner Mongolia grassland. Nature 431, 181–184.

Bardgett, R.D., Chan, K.F., 1999. Experimental evidence that soil fauna enhance nutrient mineralization and plant uptake in montane grassland ecosystems. Soil Biology & Biochemistry 31, 1007–1014.

Bardgett, R.D., Yeates, G.W., Anderson, J.M., 2005. Patterns and determinants of soil biological diversity. In: Bardgett, R.D., Usher, M.B., Hopkins, D.W. (Eds.), Biological Diversity and Function in Soils. Cambridge University Press, Cambridge, pp. 100–118.

Bardgett, R.D., Wardle, D.A., 2010. Aboveground–Belowground Linkages, Biotic Interactions, Ecosystem Processes, and Global Change. Oxford Series in Ecology and Evolution. Oxford University Press, New York.

Bardgett, R.D., van der Putten, W.H., 2014. Belowground biodiversity and ecosystem functioning. Nature 515, 505–511.

Bezemer, T.M., Lawson, C.S., Hedlund, K., Edwards, A.R., Brook, A.J., Igual, J.M., Mortimer, S.R., van der Putten, W.H., 2006. Plant species and functional group effects on abiotic and microbial soil properties and plant–soil feedback responses in two grasslands. Journal of Ecology 94, 893–904.

Bezemer, T.M., Fountain, M.T., Barea, J.M., Christensen, S., Dekker, S.C., Duyts, H., van Hal, R., Harvey, J.A., Hedlund, K., Maraun, M., Mikola, J., Mladenov, A.G., Robin, C., De Ruiter, P., Scheu, S., Setälä, H., Šmilauer, P., van der Putten, W.H., 2010. Divergent composition but similar function of soil food webs of individual plants: plant species and community effects. Ecology 91, 3027–3036.

Chen, Q.S., Li, L.H., Han, X.G., Yan, Z.D., Wang, Y.F., Yuan, Z.Y., 2003. Influence of temperature and soil moisture on soil respiration of a degraded steppe community in the Xilin River Basin of Inner Mongolia. Chinese Journal of Plant Ecology 27, 202–209.

Chen, D.M., Cheng, J.H., Chu, P.F., Hu, S.J., Xie, Y.C., Tuvshintogtokh, I., Bai, Y.F., 2015. Regional-scale patterns of soil microbes and nematodes across grasslands on the Mongolian plateau: relationships with climate, soil, and plants. Ecography 38, 622–631.

Coleman, D.C., Crossley, D.A., Hendrix, P.F., 2004. Fundamentals of Soil Ecology. Elsevier Academic Press.

Decaëns, T., 2010. Macroecological patterns in soil communities. Global Ecology and Biogeography 19, 287–302.

Fan, J.W., Zhong, H.P., Haris, W., Yu, G.Y., Wang, S.Q., Hu, Z.M., Yue, Y.Z., 2008. Carbon storage in the grasslands of China based on field measurements of above- and below-ground biomass. Climatic Change 86, 375–396.

Fang, J.Y., Yang, Y.H., Ma, W.H., Mohammat, A., Shen, H.H., 2010. Ecosystem carbon stocks and their changes in China's grasslands. Science China-Life Sciences 53, 757–765.

Feng, J., Turner, B.L., Lü, X.T., Chen, Z.H., Wei, K., Tian, J.H., Wang, C., Luo, W.T., Chen, L.J., 2016. Phosphorus transformations along a large-scale climosequence in arid and semi-arid grasslands of northern China. Global Biogeochemical Cycles 30, 1264–1275.

Harvey, G., 2001. The Forgiveness of Nature: The Story of Grass. Jonathan Cape Ltd., London.

Hou, X.Y., 1982. The Vegetation Map of the People's Republic of China (1:4000000). Sino Maps Press, Beijing.

Hu, Z.M., Fan, J.W., Zhong, H.P., Yu, G.R., 2007. Spatiotemporal dynamics of aboveground primary productivity along a precipitation gradient in Chinese temperate grassland. Science in China Series D: Earth Science 50 (5), 754–764.

Hu, Z.Z., Zhang, D.G., 2006. The Country Pasture/Forage Resources Profiles for China. http://www.fao.org/ag/AGP/AGPC/doc/Counprof/china/china1.htm.

Kang, L., Han, X.G., Zhang, Z.B., Sun, O.J., 2007. Grassland ecosystems in China: review of current knowledge and research advancement. Philosophical Transactions of the Royal Society B 362, 997–1008.

Luo, W.T., Jiang, Y., Lü, X.T., Wang, X., Li, M.H., Bai, E., Han, X.G., Xu, Z.W., 2013. Patterns of plant biomass allocation in temperate grasslands across a 2500-km transect in northern China. PLoS One 8 (8), e71749.

Luo, W.T., Elser, J.J., Lü, X.T., Wang, Z.W., Bai, E., Yan, C.F., Wang, C., Li, M.H., Zimmermann, N.E., Han, X.G., Xu, Z.W., Li, H., Wu, Y.N., Jiang, Y., 2015. Plant nutrients do not covary with soil nutrients under changing climatic conditions. Global Biogeochemical Cycles 29 (8), 1298–1308.

Ni, J., 2002. Carbon storage in grasslands of China. Journal of Arid Environments, 50, 205–218.

Nielsen, U.N., Ayres, E., Wall, D.H., Li, G., Bardgett, R.D., Wu, T.H., Garey, J.R., 2014. Global-scale patterns of assemblage structure of soil nematodes in relation to climate and ecosystem properties. Global Ecology and Biogeography 23, 968–978.

Sun, H.L. (Ed.), 2005. Ecosystems of China. Science Press, Beijing.

Suttie, J.M., Reynolds, S.G., Batello, C. (Eds.), 2005. Grassland of the World. Food & Agriculture Organization (FAO).

van der Putten, W.H., Bardgett, R.D., Bever, J.D., Bezemer, T.M., Casper, B.B., Fukami, T., Kardol, P., Klironomos, J.N., Kulmatiski, A., Schweitzer, J.A., Suding, K.N., van de Voorde, T.F.J., Wardle, D.A., 2013. Plant–soil feedback: the past, the present and future challenges. Journal of Ecology 101, 265–276.

Wang, C., Wang, X.B., Liu, D.W., Wu, H.H., Lü, X.T., Fang, Y.T., Cheng, W.X., Luo, W.T., Jiang, P., Shi, J., Yin, H.Q., Zhou, J.Z., Han, X.G., Bai, E., 2014. Aridity threshold in controlling ecosystem nitrogen cycling in arid and semi-arid grasslands. Nature Communications 5, 4799.

Wang, X.B., Lü, X.T., Yao, J., Wang, Z.W., Deng, Y., Cheng, W.X., Zhou, J.Z., Han, X.G., 2017. Habitat-specific patterns and drivers of bacterial β-diversity in China's drylands. The ISME Journal 11, 1345–1358.

Wang, X.G., Sistla, S.A., Wang, X.B., Lü, X.T., Han, X.G., 2016. Carbon and nitrogen contents in particle-size fractions of topsoil along a 3000 km aridity gradient in grasslands of northern China. Biogeosciences 13, 3635–3646.

Wardle, D.A., 2002. Communities and Ecosystems: Linking the Aboveground and Belowground Components (MPB-34). Princeton University Press, Princeton.

Wardle, D.A., Bardgett, R.D., Klironomos, J.N., Setala, H., van der Puttern, W.H., Wall, D.H., 2004. Ecological linkages between aboveground and belowground biota. Science 304, 1629–1633.

Wilson, E.O., Knopt, A.A., 2002. The future of life: the solution. Skeptic 95, 2–8.

Woodward, F.I., Lomas, M.R., Kelly, C.K., 2004. Global climate and the distribution of plant biomes. Philosophical Transactions of the Royal Society B 359, 1465–1476.

Wu, J.G., Naeem, S., Elser, J., Bai, Y.F., Huang, J.H., Kang, L., Pan, Q.M., Wang, Q.B., Hao, S.G., Han, X.G., 2015. Testing biodiversity-ecosystem functioning relationship in the world's largest grassland: overview of the IMGRE project. Landscape Ecology 30, 1723–1736.

Yang, Y.H., Fang, J.Y., Ji, C.J., Datta, A., Li, P., Ma, W.H., Mohammat, A., Shen, H.H., Hu, H.F., Knapp, B.O., Smith, P., 2014. Stoichiometric shifts in surface soils over broad geographical scales: evidence from China's grasslands. Global Ecology and Biogeography 23, 947–955.

Yu, Q., Chen, Q.S., Elser, J.J., He, N.P., Wu, H.H., Zhang, G.M., Wu, J.G., Bai, Y.F., Han, X.G., 2010. Linking stoichiometric homoeostasis with ecosystem structure, functioning and stability. Ecology Letters 13, 1390–1399.

Chapter 2

Nematode Diversity and Distribution Along the Grassland Transect

2.1 NEMATODE COMMUNITY COMPOSITION ALONG THE GRASSLAND TRANSECT

It is well recognized that in most terrestrial ecosystems, the belowground subsystem supports a greater diversity of organisms than does the aboveground subsystem (Bardgett and Wardle, 2010; Wardle et al., 2006). However, compared with diversity aboveground, our knowledge about the diversity of soil organisms is limited owing to the high richness and complexity of species in belowground subsystems (Bardgett and van der Putten, 2014). The latitudinal and altitudinal variations in some soil invertebrates have been explored, such as oribatid mites (Maraun et al., 2007), termites (Lavelle and Spain, 2001), springtails (Ulrich and Fiera, 2009), and dung beetles (Romero-Alcaraz and Ávila, 2000), some of which have similar latitudinal distribution patterns as aboveground species. Whereas some soil organisms do not exhibit a diversity gradient, such as earthworms (Lavelle et al., 1995) and soil microorganisms (Soininen, 2012). Based on available data in the soil community composition at different scales, Bardgett and van der Putten (2014) reported that no clear relationship existed between latitude and species richness for most belowground biota. Until now, the global biogeography of soil biota has been unclear because of the paucity of data about their global distribution (Bardgett and van der Putten, 2014; Decaëns, 2010).

Nematodes are the most numerous component of the mesofauna. To date, about 20,000–25,000 species have been identified, which represent less than 2.5% of estimated nematode species in soil (Orgiazzi et al., 2016). Among known species of nematodes, they occupy a range of roles. Some are important as crop pests, such as plant parasitic nematodes; others are involved in decomposition, mineralization, and nutrient cycling, such as predatory and microbivorous nematodes (Yeates and Bongers, 1999). Although nematode diversity and community structure have been studied intensively on a small scale (Lambshead et al., 2002; Porazinska et al., 2012; Powers et al., 2009; Wasilewska, 1994; Yeates, 1996), knowledge about larger-scale patterns of nematode biodiversity is relatively limited (Nielsen et al., 2014; Procter, 1984), especially at the genus or species level.

Nematodes are ubiquitous in soil and occur in almost every type of ecosystem (Coleman et al., 2004). In our study along the grassland transect, the total nematode abundance ranged from fewer than 10 to more than 2000 individuals per 100 g dry soil. The highest abundance was observed in typical steppe (in sites 47 and 48) and the lowest abundance was found in desert ecosystems (in sites 16, 18, and 26) (Fig. 2.1). Among the climatic factors, temperature and rainfall regimes have important roles in shaping nematode distribution. Total nematode abundance increased with an increase in mean annual precipitation (MAP) and showed a decreasing trend with an increase in mean annual temperature (MAT). It seems that cold and wet conditions in typical or meadow steppe ecosystems are linked to a greater abundance of total nematodes, whereas relatively dry conditions in deserts or desert steppes are linked to a lower abundance of total nematodes. In total, 32 nematode families were observed along the grassland transect. Some genera with lower frequencies were omitted from this chapter (fewer than two genera and a total nematode abundance fewer than five individuals per 100 g dry soil for each site); however, their detailed taxonomy at the genus level will be described in Chapter 3 for the purpose of identification and description.

Because there is evidence that the richness of nematode species is closely correlated to that at the genus level (Háněk and Čerevková, 2010), we present the distribution pattern of soil nematodes at the genus level. In total, 25 nematode families including 42 genera are represented for distribution. Of the 42 genera identified along the grassland transect, six families, including 12 genera belong to the bacterivores; seven families, including seven genera, belong to the fungivores; eight families, including 13 genera, are plant parasites; and five families, including 10 genera, belong to the omnivores-predators. Among the different nematode trophic groups, bacterivores are the most abundant, with an average relative abundance higher than 50%; next are the plant parasites, with an average relative abundance about 24%. For each of the other two trophic groups, fungivores and omnivores-predators, their average relative abundance is around 10% of the total nematode assemblage (Table 2.1).

Cephalobidae and Dolichodoridae are the most abundant families in the nematode group, with an average relative abundance of 53.4% and 17.6%, respectively. Among these two dominant families, *Dolichorhynchus* sp. (Dolichodoridae), belonging to plant parasites, was the most abundant genus. It had an average relative abundance of 17.6% and was represented in 41 of 62 sampling sites. The other nematode genera, with an average relative abundance higher than 5%, all belonged to the family Cephalobidae, as represented by *Cervidellus* sp. (14.1% present in 51 sites), *Acromoldavicus* sp. (13.7% present in 23 sites), *Acrobeles* sp. (9.8% present in 47 sites), and *Acrobeloides* sp. (8.6% present in 51 sites). Most of these genera were common in most sampling sites, except for *Acromoldavicus* sp., which was observed only in the

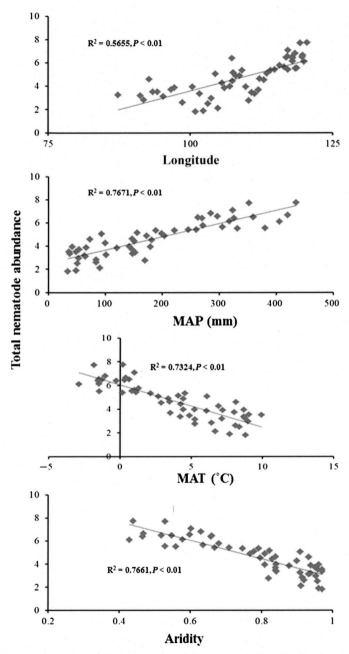

FIGURE 2.1 Total nematode abundance in relation to longitude and climatic conditions. Abundances were ln $(x+1)$ transformed.

TABLE 2.1 Relative Abundance of Nematode Genera and Their Occurrence Along the Grassland Transect

Family	Genus	Guild	Relative Abundance (%)	Occurrences in Sites
Alaimidae	Alaimus	Ba4	0.21	20
Cephalobidae	Acrobeles	Ba2	9.81	47
	Acrobeloides	Ba2	8.56	51
	Acromoldavicus	Ba2	13.68	23
	Cephalobus	Ba2	1.98	33
	Cervidellus	Ba2	14.10	51
	Chiloplacus	Ba2	2.49	42
	Eucephalobus	Ba2	2.73	44
Plectidae	Plectus	Ba2	0.34	19
	Wilsonema	Ba2	0.27	12
Prismatolaimidae	Prismatolaimus	Ba3	0.36	13
Rhabdolaimidae	Udonchus	Ba3	0.32	17
Rhabditidae		Ba1	0.10	15
Anguinidae	Ditylenchus	Fu2	2.53	41
Aphelenchidae	Aphelenchus	Fu2	1.60	26
Aphelenchoididae	Aphelenchoides	Fu2	2.45	46
Diphtherophoridae	Diphtherophora	Fu3	0.11	14
Mydonomidae	Dorylaimoides	Fu4	0.05	10
Tylencholaimidae	Tylencholaimus	Fu4	3.13	41
Tylenchidae	Filenchus	Fu2	1.94	41
	Boleodorus	Pl2	0.54	21
	Basiria	Pl2	0.22	12
	Malenchus	Pl2	0.33	18
	Tylenchus	Pl2	0.40	18
Paratylenchidae	Paratylenchus	Pl2	0.40	28
Dolichodoridae	Dolichorhynchus	Pl3	17.63	47

TABLE 2.1 Relative Abundance of Nematode Genera and Their Occurrence Along the Grassland Transect—cont'd

Family	Genus	Guild	Relative Abundance (%)	Occurrences in Sites
Hoplolaimidae	*Helicotylenchus*	Pl3	1.52	28
	Rotylenchoides	Pl3	1.73	21
	Rotylenchus	Pl3	0.27	24
Pratylenchidae	*Pratylenchus*	Pl3	0.19	15
Criconematidae	*Macroposthonia*	Pl3	0.30	22
Longidoridae	*Longidorus*	Pl5	0.66	19
Xiphinematidae	*Xiphinema*	Pl5	0.17	18
Aporcelaimidae	*Aporcelaimellus*	Om5	0.27	19
Belondiridae	*Axonchium*	Om5	0.04	10
	Dorylaimellus	Ca5	1.51	37
Campydoridae	*Campydora*	Om4	0.43	24
Nygolaimidae	*Nygolaimus*	Ca5	0.45	9
Qudsianematidae	*Crassolabium*	Om4	3.39	47
	Discolaimus	Ca5	0.16	17
	Ecumenicus	Om4	1.07	39
	Epidorylaimus	Om4	0.32	18
	Eudorylaimus	Om4	1.10	37

desert and partially in the desert steppe. Thus, except for some genera such as *Acromoldavicus* sp., Cephalobidae appear to be a cosmopolitan family; this nematode family was also identified in the other six continents across a latitudinal gradient research representing multiple ecosystem types (Nielsen et al., 2014). Other nematode families mainly showed restricted distributions along the grassland transect with relatively lower abundance, for example, Hoplolaimidae and Plectidae.

Unconstrained gradient analysis [principal component analysis (PCA)] of the distribution of nematode genera suggested that nematode distribution patterns were influenced by the type of ecosystem. In

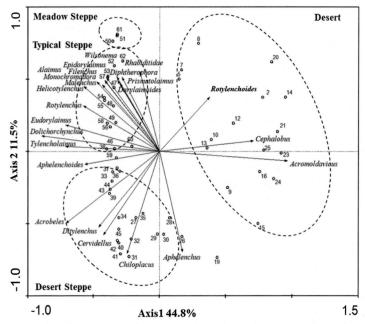

FIGURE 2.2   Principal component analysis of nematode distribution along the grassland transect. The relative abundances of nematode genera were square root transformed before the analysis; the analysis was conducted with Canoco 4.5 software.

the desert ecosystem, relatively few nematodes were present, including *Acromoldavicus*, *Cephalobus*, and *Rotylenchoides*. However, with an increase in precipitation from desert to desert steppe and to typical steppe, nematode generic richness increased. In the desert steppe, mainly fungivores (*Ditylenchus*, *Aphelenchoides*, and *Aphelenchus*) and bacterivores (*Cervidellus*, *Acrobeles*, and *Chiloplacus*) were present. In typical and meadow steppes, however, because soil and climatic conditions become more favorable for plant growth and nematode survival, some plant parasites such as *Malenchus*, *Helicotylenchus*, and *Rotylenchus*, and omnivores-predators such as *Epidorylaimus* and *Eudorylaimus* are also present in these sites. The nematode community compositions are relatively similar in typical and meadow steppes (sites 50, 51, and 61) (Fig. 2.2).

2.2 PATTERNS OF CLIMATIC DISTRIBUTION OF NEMATODE TROPHIC GROUPS

Compared with microbes, the key attribute of soil nematodes is the relation between the morphology of their feeding apparatus and their function. Nematodes feed on a variety of foods. The feeding habits of most soil nematodes can be determined by observing their anterior structures (stoma or mouth) under a microscope. Although the feeding habits of some nematodes are still poorly known, to understand the role of nematodes in soil ecosystems better, Yeates et al. (1993) suggested classifying nematodes into eight broad trophic groups according to their presumed feeding types, including: (1) plant feeders, (2) hyphal feeders, (3) bacterial feeders, (4) substrate ingesters, (5) predators, (6) unicellular eukaryote feeders, (7) dispersal- or infective-stage parasites, and (8) omnivores. In a terrestrial ecosystem, only four are generally recognized, including bacterivores, fungivores, plant parasites, and omnivores-predators (Coleman et al., 2004). Because most nematodes are active in soil throughout the year, they can provide a comprehensive measure of the functional status of soils (Ritz and Trudgill, 1999). The occurrence and abundance of nematode trophic groups can reflect changes in physical conditions in the soil and the structure of soil food webs.

Along the grassland transect, with an increase in longitude, the MAP increases from 34 mm in the west region to 436 mm in the east and the MAT ranges from −2.9 to 9.4°C. The aridity ("1−precipitation/evapotranspiration") of this transect ranges from 0.43 in the typical steppe to 0.97 in the desert ecosystem (http://www.worldclim.org/). The distribution of different nematode trophic groups was obviously influenced by climatic conditions. Among these climatic parameters, MAP was the most important parameter that influenced the distribution of nematode trophic groups. Because soil nematodes primarily inhabit water films or water-filled pore spaces in soil, they are sensitive to changes in water availability in the soil matrix (Coleman et al., 2004). The abundance of different nematode trophic groups increased with an increase in MAP, whereas an opposite pattern was observed for MAT and aridity, which showed a decreasing trend with increasing aridity and MAT (Figs. 2.3 and 2.4). The average relative abundance of omnivores-predators was relatively lower (about 10%) than for the other trophic groups; it was the most sensitive group to changes in climatic conditions, with the highest correlations ($R^2 = 0.83$; $P < .01$) observed between MAP and the abundance of omnivores-predators (Fig. 2.4). In general, the influence of climatic parameters on nematode trophic groups showed similar patterns: MAP > aridity > MAT.

For the dominant genera, *Dolichorchychus* (Dolichodoridae), *Cervidellus*, *Acrobeles*, and *Acrobeloides* (Cephalobidae) (relative abundance > 5%), their distribution had a pattern similar to that of the nematode trophic groups (Figs. 2.5 and 2.6). For *Cervidellus* and *Acrobeles*, the distributions were obviously influenced by the MAT ($R^2 = 0.63$, $P < .01$; $R^2 = 0.77$, $P < .01$), whereas for the *Acrobeloides*, it seemed that the aridity and MAP were more important than the MAT for distribution. For *Dolichorchychus*, the only dominant genus of plant parasites, the distribution was apparently constrained by the MAP, MAT, and aridity. Interestingly, for *Acromoldavicus* mainly identified in the desert and desert steppe, the distribution exhibited a pattern opposite that of the nematode trophic group; its abundance was positively correlated to MAT and aridity and negatively correlated to changes in MAP (Fig. 2.7). It seems that this genus adapted to the warm and dry conditions in the relatively drier ecosystem. These

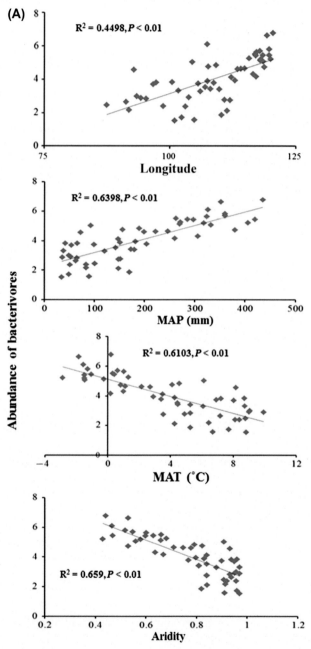

FIGURE 2.3 Variations in bacterivores (A) and fungivores (B) along the climatic gradient [longitude, mean annual precipitation (MAP), mean annual temperature (MAT), and aridity] of the grassland transect. The abundance was $\log_e$-transformed.

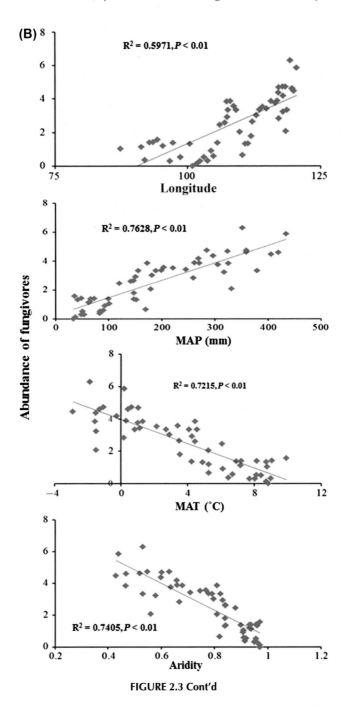

FIGURE 2.3 Cont'd

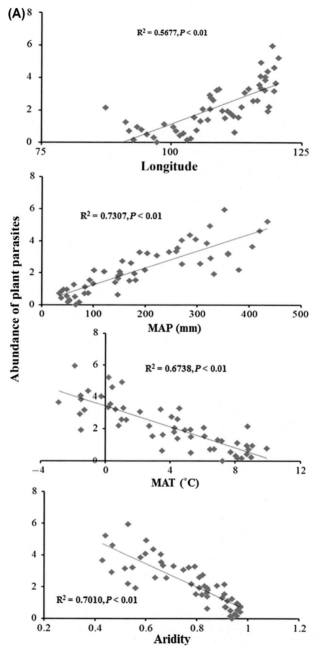

FIGURE 2.4   Variations in plant parasites (A) and omnivores-predators (B) along the climatic gradient [longitude, mean annual precipitation (MAP), mean annual temperature (MAT), and aridity] of the grassland transect. The abundance was $\log_e$-transformed.

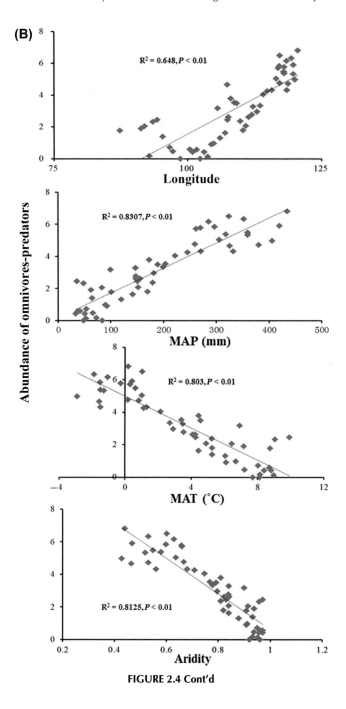

FIGURE 2.4 Cont'd

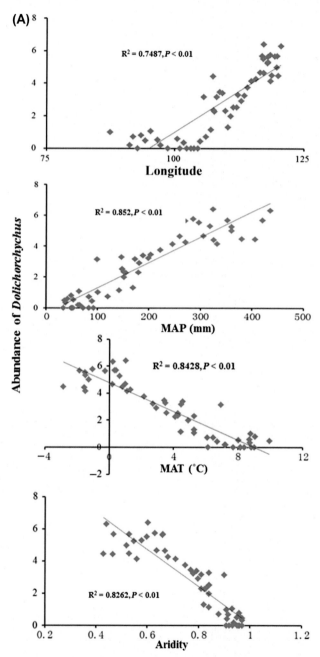

FIGURE 2.5 Variations in dominant genera *Dolichorchychus* (A) and *Cervidellus* (B) along the climatic gradient [longitude, mean annual precipitation (MAP), mean annual temperature (MAT), and aridity] of the grassland transect. The abundance was log$_e$-transformed.

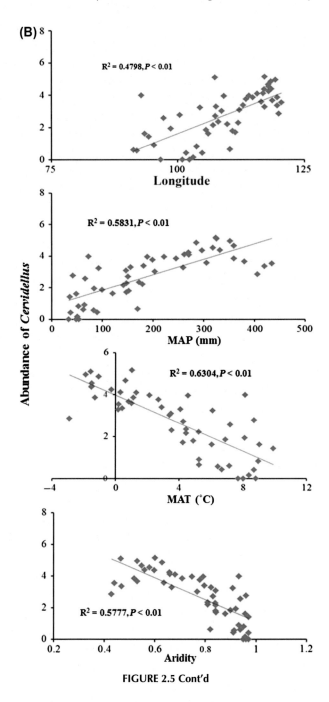

FIGURE 2.5 Cont'd

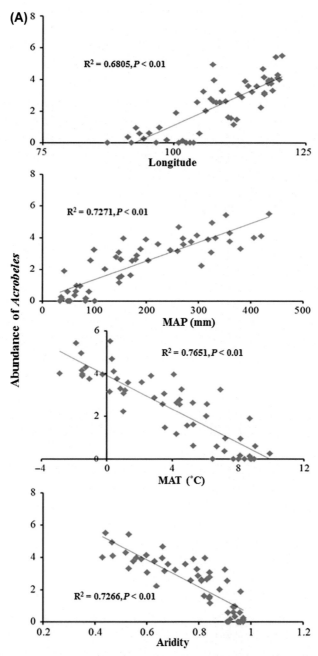

FIGURE 2.6   Variations in dominant genera *Acrobeles* (A) and *Acrobeloides* (B) along the climatic gradient [longitude, mean annual precipitation (MAP), mean annual temperature (MAT), and aridity] of the grassland transect. The abundance was $\log_e$-transformed.

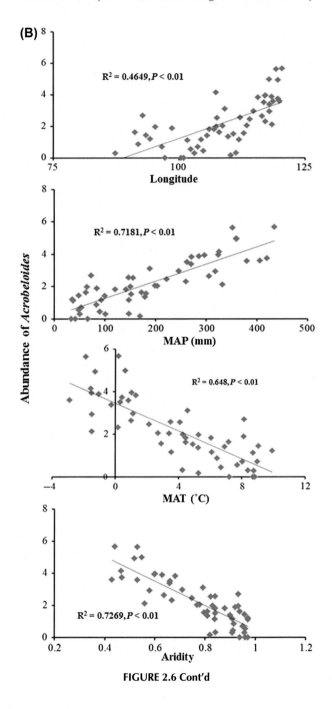

FIGURE 2.6 Cont'd

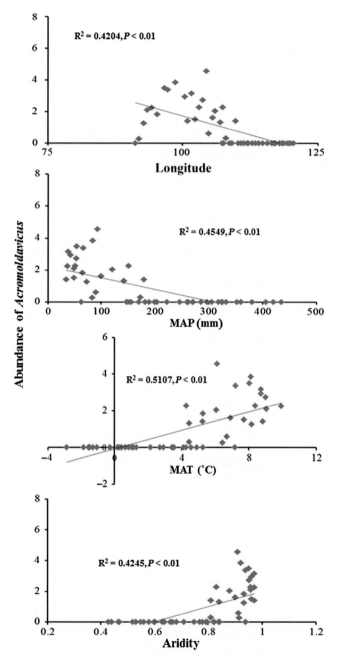

FIGURE 2.7   Variations in dominant genus *Acromoldavicus* along the climatic gradient [longitude, mean annual precipitation (MAP), mean annual temperature (MAT), and aridity] of the grassland transect. The abundance was log$_e$-transformed.

results further confirmed that research at the genus or species level is important for a better understanding of the potential mechanisms that influenced nematode distribution owing to variations in the life strategies and feeding preferences within each nematode family (Yeates et al., 1993).

2.3 NEMATODE GENERIC RICHNESS AND DIVERSITY

When viewed in relation to the conditions of an ecosystem, biodiversity is not only a matter of the number of species; it also concerns the life strategy of the constituent species. There is growing consensus that functional diversity is also important in determining the processes of an ecosystem (Bongers and Bongers, 1998). However, there is still no consistent way to quantify functional groups, which is the key to analyze the functional diversity. The structure index (SI) and enrichment index (EI) based on the integration of the "functional effect" (trophic group) and "response types" (life strategy classification) may indicate the functional diversity of soil nematodes and accelerate progress in research on diversity in nematode function (Bongers and Ferris, 1999; De Deyn et al., 2004; Ferris et al., 2001; Liang et al., 2009; Neher, 2001; Wu et al., 2002; Yeates, 2003). In this section, we present the distribution of nematode diversity in generic richness, from trophic groups to functional guilds, for a better understanding of nematode distribution patterns and the consequence they may have on ecological functioning.

Along the grassland transect, nematode richness and Shannon–Weaver diversity based on nematode genera increased with increasing MAP and decreased with increasing MAT and aridity (Figs. 2.8A and 2.9A). In contrast, nematode dominance showed an opposite trend: it decreased with increasing MAP and increased with increasing MAT and aridity (Fig. 2.8B). Trophic diversity based on the composition of different trophic groups also exhibited similar patterns with generic richness: it increased with increasing MAP and decreased with increasing MAT and aridity (Fig. 2.10A). These indices indicated that generic diversity was higher in the relatively cold and wet sites of typical and meadow steppes. In addition, nematode species composition was more proportionate and trophic structure become more complex compared with the relatively dry and warm conditions along the grassland transect. Some mechanisms have been reported to explain the distribution of soil biota, such as competition (Decaëns, 2010), resource availability (Decaëns, 2010), and plant and soil properties in soil matrix (Ettema and Wardle, 2002). However, it also depends on different scales in the study (Bardgett et al., 2005). Along the grassland transect, it seems that resource availability and climatic conditions are relatively important drivers that contribute to nematode diversity and distribution, because soil characteristics and plant biomass showed a trend similar to that of nematode diversity, such as soil C and N (negatively correlated to aridity) (Wang et al., 2016) and aboveground net primary productivity (positively correlated to MAP) (Hu et al., 2007).

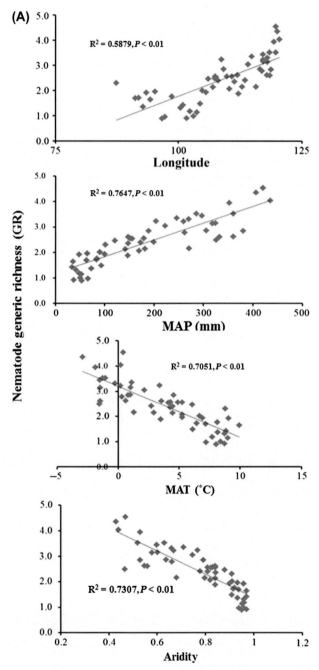

FIGURE 2.8 Variations in nematode generic richness (A) and dominance (λ) (B) along the climatic gradient [longitude, mean annual precipitation (MAP), mean annual temperature (MAT), and aridity] of the grassland transect.

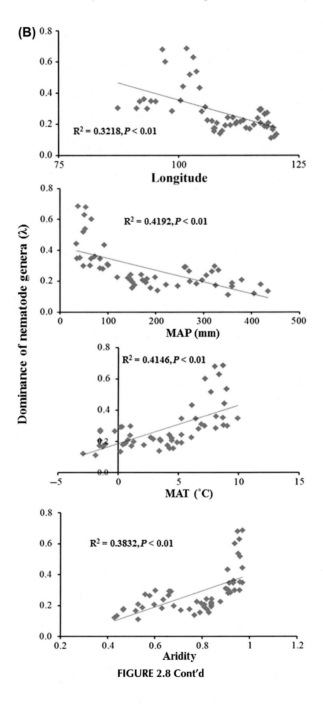

FIGURE 2.8 Cont'd

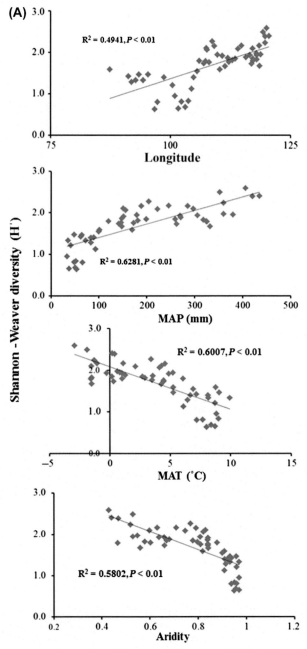

FIGURE 2.9   Variations in nematode diversity (H′) (A) and similarity (Jaccard) (B) along the climatic gradient [longitude, mean annual precipitation (MAP), mean annual temperature (MAT), and aridity] of the grassland transect.

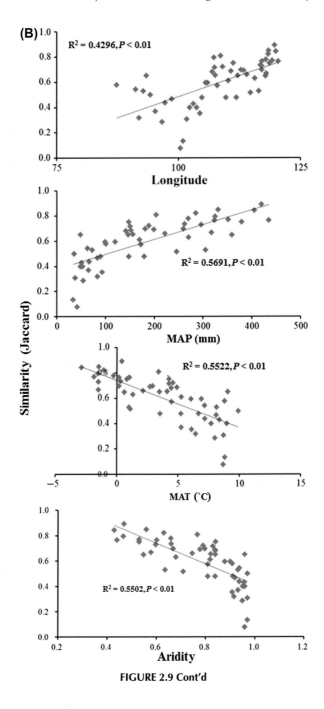

FIGURE 2.9 Cont'd

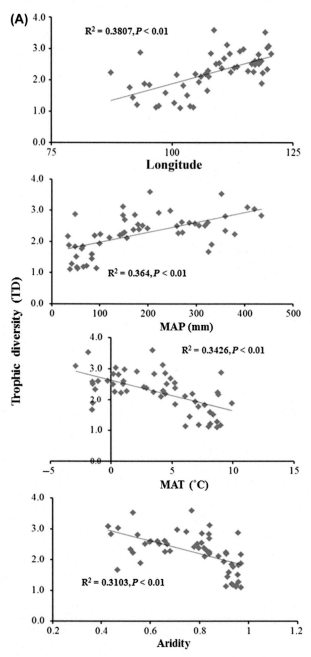

FIGURE 2.10 Changes in trophic diversity (A) and nematode channel ratio (B) along the climatic gradient [longitude, mean annual precipitation (MAP), mean annual temperature (MAT), and aridity].

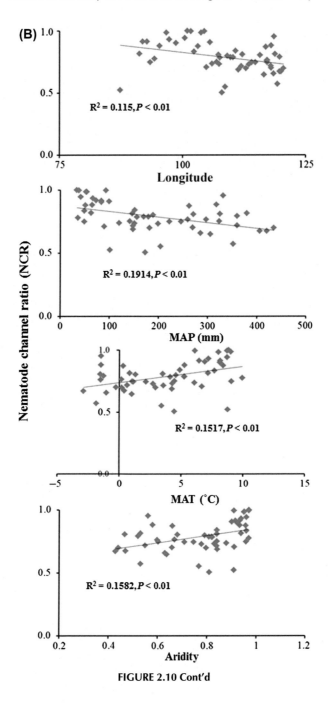

FIGURE 2.10 Cont'd

Nematode faunal analysis combining SI and EI provides a powerful tool for diagnosing the complexity of the soil food web (Wardle et al., 1995) and can be used to infer soil conditions (Ferris et al., 2001). The SI reflects the positive degree of trophic linkage in the food web and indicates the probability of regulatory effects of generalist populations on opportunist populations of soil nematodes through exploitation and competition (Ferris and Matute, 2003). Along the grassland transect, the higher SI and lower EI indicate that the soil food web was structured with a low or moderate disturbance in the meadow or typical steppe (SI > 50), and in the desert or desert steppe the soil food web (SI < 50) was degraded with stressed disturbance owing to relatively harsher climatic conditions (Fig. 2.11A and B). Because the EI indicates the abundance and activity of primary detrital consumers such as bacteria or fungi (Ferris et al., 2001), it was revealed that the supply side inputs on nematode abundance and food web function were relatively lower (EI < 25) along the grassland transect, with the bacterial pathway dominant in the decomposition (indicated by a nematode channel ratio >0.5) (Fig. 2.10B).

Beta diversity is a measure of changes in species composition along an environmental gradient (Whittaker, 1972). Not only can it reveal species distribution patterns on a relatively large spatial scale, it can also reflect changes in nematode community structure (Anderson et al., 2011). Along the grassland transect, the Jaccard index (similarity) increased with increasing MAP and decreased with increasing MAT and aridity (Fig. 2.9B), which indicated that the rate of change in nematode community composition decreased with increasing MAP and increased with increasing MAT and aridity. When the soil environment is suitable for plant growth and nematode survival, such as in typical and meadow steppes, the composition of the nematode community is more similar. These results were confirmed by PCA, in which nematode composition in typical and meadow steppes was similar, whereas nematode composition in desert and desert steppes was obviously different from that of the typical and meadow steppes (Fig. 2.2).

Although the effects of geographic location and climatic factors had important impacts on the diversity and distribution of soil nematodes, our results further indicated that the integration of taxonomic diversity and functional diversity proved to be preferable methods for indicating soil environmental conditions and the soil food web structure than did particular nematode species or ecological indices. In addition, nematode functional guilds are closely correlated to food resources and ecosystem processes, such as productivity (Ferris et al., 2001), nutrient cycling (Ferris et al., 1998), and the decomposition pathway (Freckman and Ettema, 1993). Therefore, it could be deduced from the current study that changes in the soil biotic community along the grassland transect would ultimately influence ecosystem functioning and services through changing the soil food web structure and interactions between the aboveground and belowground subsystems.

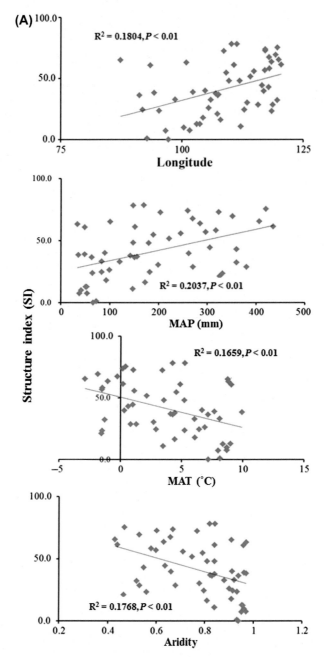

FIGURE 2.11 Changes in structure index (A) and enrichment index (B) along the climatic gradient [longitude, mean annual precipitation (MAP), mean annual temperature (MAT), and aridity].

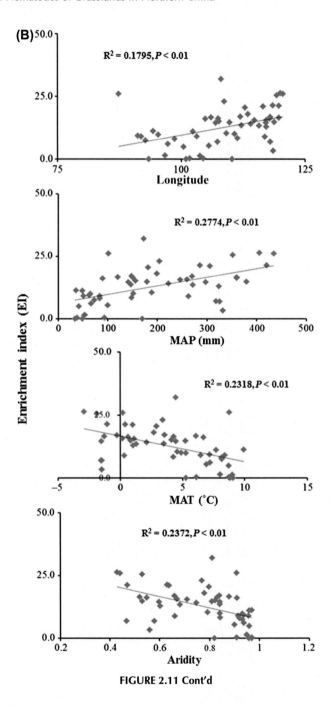

FIGURE 2.11 Cont'd

2.4 RELATIONS OF NEMATODE ASSEMBLAGE COMPOSITION WITH PLANT AND SOIL PROPERTIES

Although climatic conditions have important impacts on nematode assemblages, their distribution is also closely related to the aboveground vegetation and soil properties. Along the grassland transect, the primary productivity ranged from $<10\,g\,m^{-2}\,year^{-1}$ in the desert ecosystem to $>400\,g\,m^{-2}\,year^{-1}$ in the meadow steppe (Wang et al., 2016). In addition, soil properties varied in different ecosystems: the contents of soil total C (ranging from 2.71 to $50.33\,g\,C\,kg^{-1}$) and N (ranging from 0.14 to $4.75\,g\,N\,kg^{-1}$) decreased with increasing aridity (Wang et al., 2016). Soil pH varied from 6.46 in a typical steppe to 9.24 in a desert ecosystem. Changes in aboveground vegetation and soil conditions can directly or indirectly influence the distribution of soil nematodes.

Constrained ordination analysis indicated that soil properties and plant parameters could significantly explain the amount of variation in nematode composition along the grassland transect. The soil properties we tested could explain 18.1% of the variation, with soil pH and soil P accounting for 9% and 5%, respectively, of variation in nematode composition. Plant parameters were the basis for only a small but significant amount of variation in nematode composition (8%), with plant biomass explaining 5% of variation. Altogether, plant parameters and soil properties could account for 22.7% of variation in nematode composition.

Previous research reported that soil pH is the most important environmental factor that contributes to the biogeographic distribution of the soil microbial community. For example, on the continental scale, soil pH was a good predictor for bacterial community (Fierer and Jackson, 2006). In the current study, for nematode composition, soil pH was also an important driver that could explain 8% of variation in nematode distribution, although the content of soil C and N subtly influenced nematode distribution along the grassland transect (Table 2.2).

In our study, the content of soil P significantly influenced the distribution of soil nematodes. Phosphorus has been suggested to be a more limiting element for aboveground subsystems compared with nitrogen (Yang et al., 2014); it is nonrenewable and relatively easily depleted from the ecosystem or bonded in unavailable forms (Wardle et al., 2004b). The decrease in available P may result in an increase in the N:P ratio in soil and plant litter (Feng et al., 2016; Yang et al., 2014), influence plant productivity and the decomposition process (Wardle et al., 2004a), and then indirectly affect the belowground subsystem.

Compared with soil properties, plant parameters explained only a small amount of variation in nematode distribution. Plant biomass, not richness, significantly explained the variation in nematode composition. To date, the effects of plant biomass or net primary production on the soil food web have been inconsistent. For example, De Deyn et al. (2004) found little relation between the abundance of soil nematodes and plant shoot biomass, whereas Wardle et al. (2003)

TABLE 2.2 Constrained Ordination Analysis of Soil and Plant Characteristics on Nematode Community Composition

Explanatory Variables	Nematode Community Composition		
	% Explained	F Value	P Value
Soil parameters	**18.1**	**2.73**	**.002**
pH	**9**	**5.24**	**.002**
Total C	2	1.33	.19
Total N	2	1.44	.10
Total P	**5**	**2.60**	**.008**
Plant parameters	**8**	**2.23**	**.004**
Plant richness	2	1.05	.36
Plant biomass	**6**	**3.41**	**.002**
Plant plus soil	**22.7**	**2.31**	**.002**

P values are based on a Monte–Carlo permutation test with 999 permutations. Explained variance is based on the sum of all canonical eigenvalues. Values with significant differences are shown in bold.

reported that the biomass of microflora (the basal trophic level of the soil food web) was greatest in the most productive plant species. For a better understanding of aboveground and belowground linkages and how they influence the processes and properties of the ecosystem, further research should explore the effects of plant traits on soil communities and how this feedback in turn influences the succession and productivity of the aboveground subsystem.

REFERENCES

Anderson, M.J., Crist, T.O., Chase, J.M., Vellend, M., Inouye, B.D., Freestone, A.L., Sanders, N.J., Cornell, H.V., Comita, L.S., Davies, K.F., Harrison, S.P., Kraft, N.J.B., Stegen, J.C., Swenson, N.G., 2011. Navigating the multiple meanings of diversity: a roadmap for the practicing ecologist. Ecology Letters 14, 19–28.

Bardgett, R.D., Wardle, D.A., 2010. Aboveground–Belowground Linkages, Biotic Interactions, Ecosystem Processes, and Global Change. Oxford Series in Ecology and Evolution. Oxford University Press, New York.

Bardgett, R.D., Yeates, G.W., Anderson, J.M., 2005. Patterns and determinants of soil biological diversity. In: Bardgett, R.D., Usher, M.B., Hopkins, D.W. (Eds.), Biological Diversity and Function in Soils. Cambridge University Press, Cambridge, pp. 100–118.

Bardgett, R.D., van der Putten, W.H., 2014. Belowground biodiversity and ecosystem functioning. Nature 515, 505–511.

Bongers, T., Bongers, M., 1998. Functional diversity of nematodes. Applied Soil Ecology 10, 239–251.

Bongers, T., Ferris, H., 1999. Nematode community structure as a bioindicator in environmental monitoring. Trends in Ecology and Evolution 14, 224–228.

Coleman, D.C., Crossley, D.A., Hendrix, P.F. (Eds.), 2004. Fundamentals of Soil Ecology. Elsevier Academic Press, San Diego.

Decaëns, T., 2010. Macroecological patterns in soil communities. Global Ecology and Biogeography 19, 287–302.

De Deyn, G.B., Raaijmakers, C.E., van Ruijven, J., Berendse, F., van der Putten, W.H., 2004. Plant species identity and diversity effects on different trophic levels of nematodes in the soil food web. Oikos 106, 576–586.

Ettema, C.H., Wardle, D.A., 2002. Spatial soil ecology. Trends in Ecology and Evolution 17, 177–183.

Feng, J., Turner, B.L., Lü, X.T., Chen, Z.H., Wei, K., Tian, J.H., Wang, C., Luo, W.T., Chen, L.J., 2016. Phosphorus transformations along a large-scale climosequence in arid and semi-arid grasslands of northern China. Global Biogeochemical Cycles 30, 1264–1275.

Ferris, H., Matute, M.M., 2003. Structural and functional succession in the nematode fauna of a soil food web. Applied Soil Ecology 23, 93–110.

Ferris, H., Venette, R.C., van der Meulen, H.R., Lau, S.S., 1998. Nitrogen mineralization by bacterial-feeding nematodes: verification and measurement. Plant and Soil 203, 159–171.

Ferris, H., Bongers, T., de Goede, R.G.M., 2001. A framework for soil food web diagnostics: extension of the nematode faunal analysis concept. Applied Soil Ecology 18, 13–29.

Fierer, N., Jackson, R.B., 2006. The diversity and biogeography of soil bacterial communities. Proceedings of the National Academy of Sciences of the United States of America 103, 626–631.

Freckman, D.W., Ettema, C.H., 1993. Assessing nematode communities in agroecosysems of varying human intervention. Agricuture, Ecosystems and Environment 45, 239–261.

Háněk, L., Čerevková, A., 2010. Species and genera of soil nematodes in forest ecosystems of the Vihorlat Protected Landscape Area, Slovakia. Helminthologia 47, 123–135.

Hu, Z.M., Fang, J.W., Zhong, H.P., Yu, G.R., 2007. Spatiotemporal dynamics of aboveground primary productivity along a precipitation gradient in Chinese temperate grassland. Science in China Series D: Earth Sciences 50, 754–764.

Lambshead, P.J.D., Brown, C.J., Ferrero, T.J., Mitchell, N.J., Smith, C.R., Hawkins, L.E., Tietjen, J., 2002. Latitudinal diversity patterns of deep-sea marine nematodes and organic fluxes: a test from the central equatorial Pacific. Marine Ecology Progress Series, 236, 129–135.

Lavelle, P., Lattaud, C., Trigo, D., Barois, I., 1995. Mutualism and biodiversity in soils. Plant and Soil 170, 23–33.

Lavelle, P., Spain, A.V. (Eds.), 2001. Soil Ecology. Kluwer Academic Publishers, Dordrecht.

Liang, W.J., Lou, Y.L., Li, Q., Zhong, S., Zhang, X.K., Wang, J.K., 2009. Nematode faunal response to long-term application of nitrogen fertilizer and organic manure in Northeast China. Soil Biology and Biochemistry 41, 883–890.

Maraun, M., Schatz, H., Scheu, S., 2007. Awesome or ordinary? Global diversity patterns of oribatid mites. Ecography 30, 209–216.

Neher, D.A., 2001. Role of nematodes in soil health and their use as indicators. Journal of Nematology 33, 161–168.

Nielsen, U.N., Ayres, E., Wall, D.H., Li, G., Bardgett, R.D., Wu, T.H., Garey, J.R., 2014. Global-scale patterns of assemblage structure of soil nematodes in relation to climate and ecosystem properties. Global Ecology and Biogeography, 23, 968–978.

Orgiazzi, A., Bardgett, R.D., Barrios, E., Behan-Pelletier, V., Briones, M.J.I., Chotte, J.L., De Deyn, G.B., Eggleton, P., Fierer, N., Fraser, T., Hedlund, K., Jeffery, S., Johnson, N.C., Jones, A., Kandeler, E., Kaneko, N., Lavelle, P., Lemanceau, P., Miko, L., Montanarella, L., Moreira, F.M.S., Ramirez, K.S., Scheu, S., Singh, B.K., Six, J., van der Putten, W.H., Wall, D.H. (Eds.), 2016. Global Soil Biodiversity Atlas. European Commission, Publications Office of the European Union (Luxembourg).

Porazinska, D.L., Giblin-Davis, R.M., Powers, T.O., Thomas, W.K., 2012. Nematode spatial and ecological patterns from tropical and temperate rainforests. PLoS ONE 7 (9), e44641.

Powers, T.O., Neher, D.A., Mullin, P., Esquivel, A., Giblin-Davis, R.M., Kanzaki, N., Stock, S.P., Mora, M.M., Uribe-Lorio, L., 2009. Tropical nematode diversity: vertical stratification of nematode communities in a Costa Rican humid lowland rainforest. Molecular Ecology 18, 985–996.

Procter, D.L.C., 1984. Towards a biogeography of free-living soil nematodes. I. Changing species richness, diversity and densities with changing latitude. Journal of Biogeography 11, 103–117.

Ritz, K., Trudgill, D.L., 1999. Utility of nematode community analysis as an integrated measure of the functional state of soils: perspective and challenges. Plant and Soil 212, 1–11.

Romero-Alcaraz, R., Ávila, J.M., 2000. Effect of elevation and type of habitat on the abundance and diversity of scarabaeoid dung beetle (carabaeoidea) assemblages in a Mediterranean area from Southern Iberian Peninsula. Zoological Studies 39, 351–359.

Soininen, J., 2012. Macroecology of unicellular organisms – patterns and processes. Environmental Microbiology Reports 4, 10–22.

Ulrich, W., Fiera, C., 2009. Environmental correlates of species richness of European springtails (Hexapoda: Collembola). Acta Oecologia 35, 45–52.

Wang, X.G., Sistla, S.A., Wang, X.B., Lü, X.T., Han, X.G., 2016. Carbon and nitrogen contents in particle-size fractions of topsoil along a 3000 km aridity gradient in grasslands of northern China. Biogeosciences 13, 3635–3646.

Wardle, D.A., Yeates, G.W., Watson, R.N., Nicholson, K.S., 1995. The detritus food web and the diversity of soil fauna as indicators of disturbance regimes in agroecosystems. Plant and Soil 170, 35–43.

Wardle, D.A., Yeates, G.W., Williamson, W., Bonner, K.I., 2003. The response of a three trophic level soil food web to the identity and diversity of plant species and functional groups. Oikos 102, 45–56.

Wardle, D.A., Bardgett, R.D., Klironomos, J.N., Setälä, H., van der Putten, W.H., Wall, D.H., 2004a. Ecological linkages between aboveground and belowground biota. Science 304, 1629–1633.

Wardle, D.A., Walker, L.R., Bardgett, R.D., 2004b. Ecosystem properties and forest decline in contrasting long-term chronosequences. Science 305, 509–513.

Wardle, D.A., Yeates, G.W., Barker, G.M., Bonner, K.I., 2006. The influence of plant litter diversity on decomposer abundance and diversity. Soil Biology and Biochemistry 38, 1052–1062.

Wasilewska, L., 1994. The effect of age of meadows on succession and diversity in soil nematode communities. Pedobiologia 38, 1–11.

Whittaker, R.H., 1972. Evolution and measurement of species diversity. Taxon 21, 213–251.

Wu, J.H., Fu, C.Z., Chen, S.S., Chen, J.K., 2002. Soil faunal response to land use: effect of estuarine tideland reclamation on nematode communities. Applied Soil Ecology 21, 131–147.

Yang, Y.H., Fang, J.Y., Ji, C.J., Datta, A., Li, P., Ma, W.H., Mohammat, A., Shen, H.H., Hu, H.F., Knapp, B.O., Smith, P., 2014. Stoichiometric shifts in surface soils over broad geographical scales: evidence from China's grasslands. Global Ecology and Biogeography 23, 947–955.

Yeates, G.W., 1996. Diversity of nematode faunae under three vegetation types on a pallic soil in Otago, New Zealand. New Zealand Journal of Zoology 23, 401–407.

Yeates, G.W., 2003. Nematodes as soil indicators: functional and biodiversity aspects. Biology and Fertility of Soils 37, 199–210.

Yeates, G.W., Bongers, T., 1999. Nematode diversity in agroecosystems. Agriculture, Ecosystems and Environment 74, 113–135.

Yeates, G.W., Bongers, T., de Goede, R.G.M., Freckman, D.W., Georgieva, S.S., 1993. Feeding habits in soil nematode families and genera – an outline for soil ecologists. Journal of Nematology 25, 315–331.

Chapter 3

Nematode Genera and Species Description Along the Transect

3.1 BACKGROUND ON NEMATODE TAXONOMY

The term "nematode" is derived from two Greek words: *nema* (thread) and *eidos* (like). Nematodes are thus basically thread-like organisms. They can be defined as transparent, bilaterally symmetrical, unsegmented, pseudo-coelomate, triploblastic, thread-like animals. They are popularly known as roundworms because the cross-section of their body is circular. However, to many of us, nematodes are something unheard of or unseen because they are tiny organisms that cannot be seen with the naked eye. Besides, they also lead a hidden life either in the soil or in the body of other organisms, whether plants or animals. We cannot think of a place where they are absent. No other multicellular organisms are so diverse in their distribution of habitat. They should never be neglected because they have an important role in the economy of a country, and many of them are parasites of plants, mushrooms, domestic animals, and even humans. Every year the world loses millions of dollars as a result of these tiny organisms. Almost all crops are affected by the presence of these creatures. Many forests, orchards, golf courses, etc., become devastated owing to the presence of different plant-parasitic nematodes. On the other hand, some of them are good friends to farmers because they kill different types of insect pests and some increase the fertility of soil, hasten the process of mineralization, and help in nutrient cycling in the belowground subsystem.

3.1.1 General Information on Soil Nematodes' Life Strategies

Different Life Strategies: It is a well-accepted fact that nematodes are second only to insects in terms of species diversity. However, in terms of habitat diversity, nematodes leave all other animal groups far behind. They may have achieved this tremendous feat through their surprisingly varied lifestyles. They play almost every possible role in terms of their life strategies, reproduction, growth, food, feeding habits, etc. Besides being small and simple organisms they can adapt to different geophysicochemical conditions. Upon the onset of unfavorable conditions many can go into a different

Soil Nematodes of Grasslands in Northern China. http://dx.doi.org/10.1016/B978-0-12-813274-6.00003-X

mode of life such as *omnivory* (adaptation to other food sources on which they normally do not feed), *cryptobiosis* (survival with no detectable metabolic activity), *dormancy* (a condition of lowered metabolism), or *dauer stage* (also a kind of metabolically suppressed survival stage). Some also can withstand complete dryness on the surface of rocks. Depending on their feeding habits, *terrestrial* (soil-dwelling) and *aquatic* (living in water) nematodes are generally grouped into *microbivorous/saprophagous* (living on microbes or decaying organic materials), *herbivorous* (living on plants), and *omnivorous/predatory* (living on other small nematodes or other soil microorganisms). Life can be considered a process of eating and being eaten. Almost all nematodes derive their energy from their food. However, there are certain stages in the course of their development during which nematodes do not feed. There are also some examples in which only females feed and males are nonfeeding and also lack a functional pharynx, intestines, and so on (e.g., members of the family Criconematidae). Such males consume fat globules stored in their body and act as mates of females in sexual reproduction.

Role of Nematodes in Our Life: However small nematodes are, they occupy an important position in the food chain of the soil subsystem. Saprophagous or microbivorous nematodes have an important role in the decomposition of organic materials. Those that are plant-parasitic also affect us directly or indirectly through economic loss. When a nematode attacks a plant, the plant is first injured physically and the wound is exposed to the environment for further secondary infections by other microorganisms. Some plant-parasitic nematodes also carry viruses in their alimentary tract and the viruses are transmitted to the plants through their feeding. The annual loss of crops and other plant products as a result of nematode infestation is in the many millions of dollars. According to Handoo (1998), annual crop losses owing to nematodes approximate $80 billion. With such a fast-increasing population and simultaneously decreasing agriculturally farmable land, no country can afford such a loss. Sasser and Freckman (1987) estimated that crop losses owing to nematodes occur more in tropical and subtropical climates (14.6%) than in developed countries (8.8%). Unfortunately, only about 0.2% of the crop value lost to nematodes is used to fund nematological research.

Nematodes are not always our enemies. Some are good experimental models for various biological studies such as reproductive biology, molecular biology, and physiology. This can be attributed to their short life cycle and their prolific and easily cultivated nature, compounded by the transparency of their body cuticle. A good number of them are indicators of soil health, soil type, pH, pollution level, etc. They are also suitable for ecological studies. One can easily correlate ecological succession with the corresponding nematode population.

3.1.2 General Morphological Terminologies and Structures of Nematodes

Body Organization: The body organization of nematodes can simply be defined as a tube within a tube. The body wall represents the outer tube and the alimentary canal represents the inner tube. These two tubes join anteriorly at the mouth and posteriorly at the anus in females (the openings of the digestive and genital systems are separate in females) and the cloaca in males (open digestive and genital systems are common in males).

Body Posture: Nematodes assume a particular body posture upon death or on heat relaxation or fixation; therefore, body posture is an important diagnostic feature. Upon death, nematodes generally take on a ventral curvature of varying degrees. However, they may be almost straight or spiral, or even dorsally curved.

Body Symmetry: Except for a few nematode species such as *Bunonema*, almost all nematodes are bilaterally symmetrical externally (when the body is cut into two halves through the midsagittal line dorsoventrally). However, some internal organs show different types of symmetry, such as triradial (feeding apparatus and esophagus (pharynx)), tetraradial (hypodermal chords), and hexaradiate (lips and labial papillae and in particular the basal knobs of *Hexatilus*). Asymmetry can also be seen in organs such as the genital, nervous, and excretory systems.

Body Regions: Generally the body of the nematode is divided into different regions: lip or cephalic, stomal, esophageal or pharyngeal, cardial, prevulval, vulval or advulval, postvulval, precloacal or preanal, anal or adanal or cloacal, postanal or postcloacal, caudal, etc. (Fig. 3.1).

- The lip or cephalic region is the distance from the extreme anterior end to the base of the lips.
- The esophageal or pharyngeal region is the body distance from the base of the lips to the posterior end of the pharynx/esophagus.
- The cardial region is the region of the junction of the esophagus and intestine.
- The prevulval region denotes the distance between the base of the esophagus/pharynx to the level of the vulva.
- The vulval or advulval region denotes the area near the vulva.
- The postvulval region is the body distance between the vulva and the anus.
- The precloacal or preanal region denotes the area anterior to the cloaca or anus.
- The anal or adanal or cloacal region indicates the area near the anus or cloaca.
- The caudal region denotes the tail region.

Body Wall: The nematode body wall is made of an outer cuticle, hypodermis, and inner somatic layer. The outermost layer of the body wall is known as a

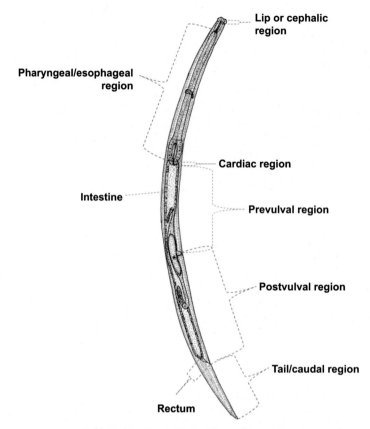

FIGURE 3.1   General body regions of nematodes.

cuticle. It is noncellular, nonliving, transparent, semipermeable, and flexible in nature. It gives shape to the body and protects the soft and vital internal organs from the external environment. It also acts as an osmotic barrier between the internal body and the external environment. During the course of development from egg to adult, the cuticle is shed at least four times. This phenomenon is known as molting. Different types of markings, ornamentations, or structures that are of taxonomic importance are associated with the cuticle.

Transverse Markings: The outermost layer of cuticle has a particular pattern of transverse striations or grooves although it appears smooth at low magnifications (Fig. 3.2A). The transverse markings may be fine or coarse, or they may further be modified into annulations forming grooves between two successive annules (Fig. 3.2C, D, and F).

Longitudinal Markings: Besides transverse markings, the cuticle may also bear longitudinal markings. Such longitudinal markings may be restricted to the

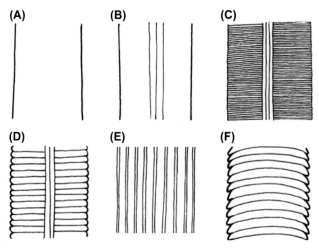

FIGURE 3.2 Different markings on cuticle (A–F). (A) Smooth-appearing cuticle at low magnification; (B) Smooth cuticle with lateral markings; (C) Cuticle with fine transverse striations; (D) Cuticle with coarse annules; (E) Cuticle with longitudinal markings; (F) Cuticle with backwardly directed annules.

lateral chords or they may be present on the entire body (Fig. 3.2E). The lateral side of the nematode body is mostly characterized by the presence of lateral lines or incisures (Chromadorea) (Fig. 3.2B), the number of which is also of some taxonomic importance for lateral hypodermal glands (for some Enoplia).

Cuticular Ornamentations: The cuticles of different species possess differently modified structures known as cuticular ornamentations. The ornamentations may be in the form of dots, warts, elevations, depressions, spines, or projections either on the whole body or restricted to the anterior or posterior region of the body (Fig. 3.2F).

Body or Somatic Setae: These are fine, setose structures present on the cuticle. They are found mostly in free-living and aquatic forms.

Body Openings: Body openings include oral openings, amphidial apertures, excretory pores, hypodermal gland openings, anal/cloacal openings, vulval openings, openings of spinneret glands, and openings of organellum dorsale. Except for the organellum dorsale and amphidial apertures, all openings are located midventrally. Amphidial apertures are present laterally on the lateral lips or posterior. Excretory pores are usually located ventrally in the midcervical region or posterior to it.

Hypodermis: This is a thin layer located just beneath the cuticle. It is characterized by four longitudinal protuberances (one dorsal, one ventral, and two lateral) toward the celomic cavity dividing the body into four sections: two sub-dorsal and two subventral. The longitudinal protuberances are also known as chords. The lateral chords are generally larger than the dorsal and the ventral chords. The ventral and the dorsal chords also contain longitudinal body nerves whereas the lateral chords carry lateral excretory canals.

Somatic Musculature: This is represented by a layer of longitudinal, spindle-shaped muscle cells attached to the hypodermis. It is specialized for the movement of both internal and external organs. Each muscle cell has two parts: sarcoplasmic and fibrillar. The sarcoplasmic part carries the nucleus and the fibrillar part carries oblique contractile fibers. Each cell is connected to the nerves of longitudinal and ventral chords. Usually somatic musculature is of less taxonomic importance.

Lip or Cephalic Region: This is one of the most important parts of the nema-tode body from a taxonomic point of view because it has a great variety of modifications in different groups of nematodes (Fig. 3.3). The lip or cephalic region may be continuous with the body or it may be set off from the body by a depression or constriction. An elevated, disc-like structure may also be associ-ated with the lip region. Other structures are associated with the cephalic region.

Lips and Labial Papillae: In nematodes, six lips surround the oral opening; hence it shows hexaradiate symmetry. The lips are arranged in such a way that two are lateral, two are subventral, and two are subdorsal. Except for the later-als, which have two papillae, each lip has three papillae. The labial papillae are arranged into two circlets: an inner and an outer circlet. In the inner circlet all lips have one papilla each and in the outer circlet all submedians have two papil-lae each, whereas the lateral lips have one papilla each. The loss of one papilla each on the lateral lips is correlated with the presence of amphidial openings and apertures on the lateral lips.

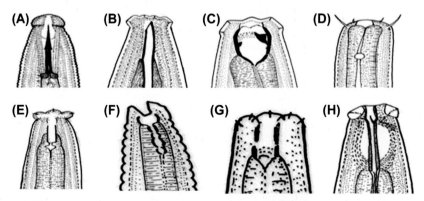

FIGURE 3.3   Lip or cephalic region of nematodes (A–H). (A) Tylenchid; (B) Dorylaimoid; (C) Mononchid; (D) Monhysterid; (E) Rhabditid; (F) Cephalobid; (G) Panagrolaimid (H) Campydoryd.

Cephalic Framework: This is a circular ring or basket-like cuticularized structure surrounding the stoma. The degree of sclerotization varies from one species to another; hence it is of taxonomic importance.

Amphids: These are a pair of innervated chemoreceptor organs located on the lateral sides of the body in the pharyngeal region. The openings of the amphids, also known as amphidial apertures, may be pore-like, circular, oval, or slit-like, and may be located on the lateral lips or posterior to them.

Deirids: These are also paired structures present in the form of protuberances on the cuticle in the midlateral fields in the pharyngeal region at approximately the level of the excretory pore. These structures are not necessarily present in all nematode groups. Although the presence or absence of deirids is of taxonomic importance, the position of these structures is not taxonomically important because the location of deirids is fairly uniform.

Phasmids: Phasmids are also paired structures present in the midlateral fields. They have been shown to be chemorepulsive in function. They may be generally located posterior to the anus or cloaca. However they may be present adanal, preanal, or further anterior. The shape and position of phasmids are of taxonomic value.

Digestive Tract/System: Most nematode groups possess a well-developed, well-defined digestive system. It is the inner tube extending from the oral opening to the anus/cloaca. The digestive system is composed of three main parts: the *stomodeum*, *mesenteron*, and *proctodeum*. The *stomodeum* consists of the stoma, esophagus, and cardia (esophago-intestinal junction). The *mesenteron*, also known as the midgut, is the intestine proper whereas the *proctodeum*, also known as the hind gut, is the rectum. The *stomodeum* is lined with cuticle and is variously shaped in different groups of nematodes; it is of high taxonomic importance, whereas the *mesenteron* and *proctodeum* are simple tubular structures and are of no taxonomic value.

Stoma: This is known by different names such as the buccal cavity, buccal capsule, or sometimes the mouth cavity. It is the anterior-most portion of the digestive system. The stoma is greatly variable in shape and size in different nematode groups and hence is of extreme importance in taxonomy. Based on the shape and size of the stoma, one can identify the feeding habit of the nematode; therefore it is important for ecological studies (Fig. 3.3). The anterior-most portion of the stoma is also known as the vestibule. The stoma may be very small or large. It may be tubular or funnel shaped, barrel shaped, or spacious. It may have no armature or have an armature that resembles tooth, teeth, warts, denticles (minute tooth-like projections), or a stylet (Fig. 3.4).

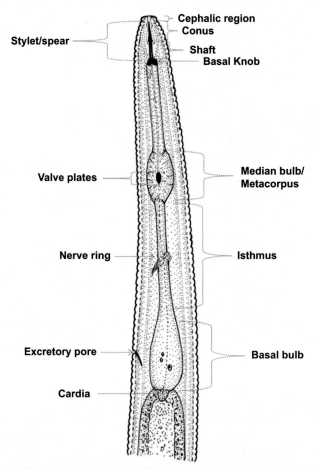

FIGURE 3.4   Different parts of the feeding apparatus in tylenchid nematodes.

All plant-parasitic nematodes have a protrusible, hypodermic needle–like stylet or spear. The stylet has an anterior conical part known as a conus and a posterior cylindrical part known as a shaft. Both the conus and shaft have a narrow lumen. Three knobs are attached posteriorly to the shaft. The conus, shaft, and basal knobs have differed to varying degrees in length, size, and shape; therefore they are of taxonomic importance (Fig. 3.4).

In most dorylaim nematodes, the stylet is known as the odontostyle, whereas in the smaller group of dorylaim called the nygolaims, the odontostyle is known as the onchiostyle because the stylet develops from a mural tooth. A circular, sclerotized ring-like structure called the guiding ring/apparatus surrounds the odontostyle. It may be single or double. The odontostyle is followed by a largely cylindrical structure called the odontophore. The length, shape, and aperture of the odontostyle vary. The odontophore may be cylindrical, simply rodlike, or

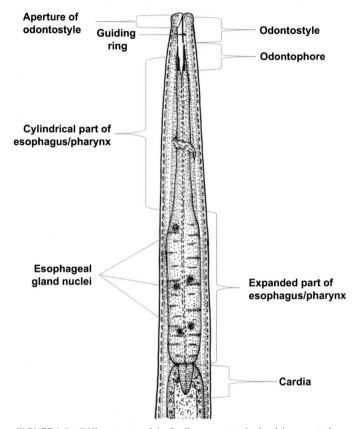

Aperture of odontostyle

Guiding ring

Odontostyle

Odontophore

Cylindrical part of esophagus/pharynx

Esophageal gland nuclei

Expanded part of esophagus/pharynx

Cardia

FIGURE 3.5   Different parts of the feeding apparatus in dorylaim nematodes.

basally flanged. Therefore both the odontostyle and odontophore are taxonomically important (Fig. 3.5).

Predatory nematodes such as mononchs have a wide buccal/stomal cavity; the stomal wall is usually provided with well-developed armatures such as dorsal tooth, subventral teeth, denticles, etc. When only one tooth is present in the buccal cavity, it is usually on the dorsal wall. Generally, the dorsal tooth is the largest one (Fig. 3.6).

Free-living nematodes, especially of the order Rhabditida, have a stoma composed of three circular, ring-like cuticularized structures: cheilostom, gymnostom, and stegostom. The anterior-most part is known as the cheilostom and is followed by the gymnostom. Neither the cheilostom nor the gymnostom is covered by pharyngeal tissue. The posterior-most part of the stoma is known as the stegostom and is always covered by pharyngeal tissue. The stegostom can be further divided into the prostom, mesostom, metastom, and telostom. Generally the prostom and mesostom are amalgamated with the gymnostom, whereas the

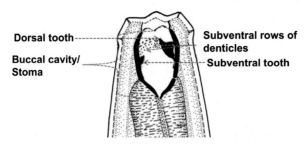

Dorsal tooth

Buccal cavity/
Stoma

Subventral rows of
denticles

Subventral tooth

FIGURE 3.6 Different parts of the feeding apparatus in mononchid nematodes.

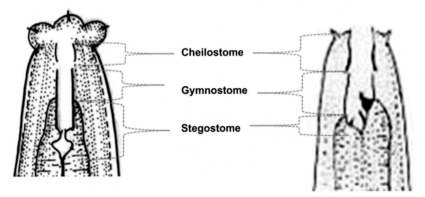

Cheilostome

Gymnostome

Stegostome

FIGURE 3.7 Different parts of the feeding apparatus in rhabditid nematodes.

metastom usually has a glottoid apparatus bearing denticles. All of the cuticu-larized ring-like structures are known as rhabdions; their corresponding names are cheilorhabdions, gymnorhabdions, etc. (Fig. 3.7).

Esophageal/Pharyngeal Glands: These are unicellular, uninucleate cells associ-ated with the pharyngeal tissue. The number of these glands varies in different groups of nematodes, ranging from three to many. The plant-parasitic tylen-chid nematodes generally have three glands; the predatory dorylaims have five. In some plant-parasitic groups the glands may form a lobe extending over the intestine. Among the dorylaim nematodes the arrangement and number of these gland nuclei are species specific and hence are of taxonomic importance (Figs. 3.8A and B).

Esophago-intestinal Junction or Cardia: This is a disc or tongue-like struc-ture connecting the esophagus and intestine (Figs. 3.4 and 3.5). It serves as a valve between the two and prevents ingested food from coming back from the intestine. The posterior part of the cardia centrally protrudes into the lumen of the intestine. Cardia may or may not be provided with one to three unicellular glands.

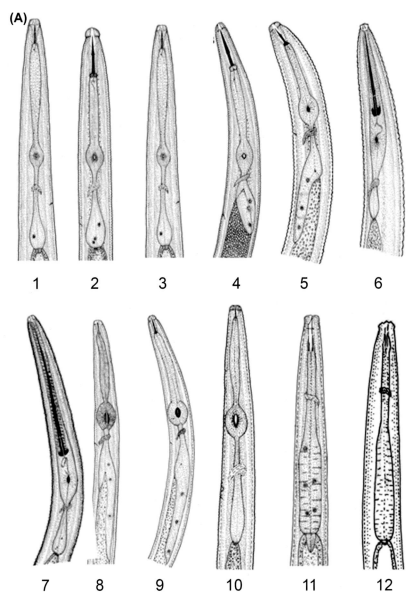

FIGURE 3.8A   Pharynx in different families: (1) Tylenchidae; (2) Dolichodoridae; (3) Psilenchidae; (4) Hoplolaimidae; (5) Pratylenchidae; (6) Criconematidae; (7) Paratylenchidae; (8) Aphelenchidae; (9) Aphelenchoididae; (10) Paraphelenchidae; (11) Dorylaimidae; (12) Aporcelaimidae.

(B)

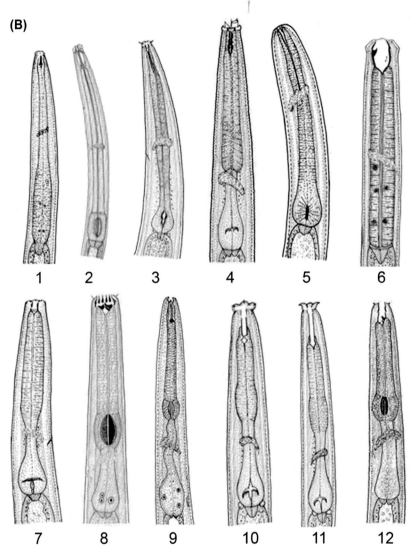

FIGURE 3.8B Pharynx in different families: (1) Tylencholaimidae; (2) Campydoridae; (3) Plectidae; (4) Cephalobidae; (5) Rhabdolaimidae; (6) Mononchidae; (7) Panagrolaimidae; (8) Neodiplogastridae; (9) Tylopharyngidae; (10) Rhabditidae; (11) Diploscapteridae; (12) Diplogastridae.

Intestine: This is the largest part in the whole digestive system. It connects the cardia anteriorly and the rectum or prerectum (in the Dorylaimida) posteriorly (Fig. 3.1). It is a tubular, one cell-thick, elongated structure. The intestinal cells are large and polygonal, or sometimes syncytial.

Prerectum: In some groups such as Dorylaimida the intestine connects posteriorly with a structure called the prerectum, which is usually different from the intestine proper in color, thickness, and the texture of the food it contains, etc. The length of the prerectum varies among species and has some taxonomic value, especially in Dorylaimida.

Rectum: This is the posterior-most part of the digestive system. It is much thinner than the intestine and connects the intestine/prerectum anteriorly and posteriorly to the anus (Fig. 3.1). It is provided with a sphincter (a circular and muscular contractile ring) at the junction with the intestine. Three unicellular glands may also be associated with the rectum anteriorly.

Anus/Cloaca: In female nematodes, both reproductive and digestive systems have separate openings to the exterior. The anus is the posterior-most part of the rectum. However, in males, the reproductive and digestive systems have a common opening to the exterior called the cloaca. The waste product of the digestive system and sperm are ejaculated through the common opening, the cloaca.

Reproductive System

Female: The female genital system is composed of the ovary, oviduct, spermatheca, uterus, vagina, and vulva. A female nematode may have only one branch of the genital tract or a pair of tracts (Fig. 3.9). When there is only one functional genital tract, it is said to be *monodelphic*, and if there is a pair of functional genital tracts, it is called *didelphic* (Fig. 3.9A). If the female has only one gonad extending posterior to the vulva, the condition is called *opisthodelphic* (Fig. 3.9B). If the female has only one gonad extending anterior to the vulva, the condition is called *prodelphic* (Fig. 3.9C). The ovaries of a female with two sets of genital tracts may be extended to the opposite sides (*amphidelphic*) (Fig. 3.9A) or on the same side, as in the case of *Meloidogyne* (*didelphic–prodelphic*) (Fig. 3.9D). If there is a reduced, nonfunctional gonad in addition to one that is fully functional, the condition is called a *pseudoprodelphic* or *pseudoopisthodelphic* gonad, depending on the anterior or posterior extension of the functional gonad from the vulva. The nonfunctional branch, either *prevulval* or *postvulval*, based on its position in relation to the vulva, may have different degrees of development. If it is only a sac, it may be called a prevulval or postvulval uterine sac (Fig. 3.9C).

Ovary: This is the distal terminal part of the female genital system and is characterized by the presence of oocytes at different developmental stages. Oocytes toward the uterus are always larger. The oocytes may be arranged singly linearly or they may be in two or more rows. When the tips of the ovaries are directed toward the body exterior, the gonad is said to be *outstretched*. If the tip of the ovary is directed toward the vulva with the flexure present in the ovary itself, it is said to be *reflexed*. However, in some nematodes there is no flexure in the ovary itself; rather, the flexure is present at the junction of the ovary and oviduct, keeping the tip of the ovary directed toward the vulva.

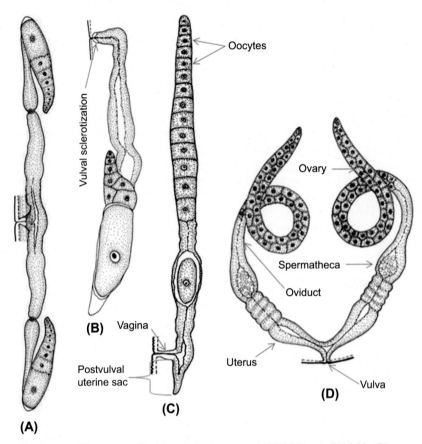

FIGURE 3.9   Different types of female reproductive system: (A) didelphic-amphidelphic; (B) mono-opisthodelphic; (C) monoprodelphic (with postvulval uterine sac); and (D) didelphic-prodelphic.

In such a situation the ovary is said to be *reversed*. The tip of the ovary is the germinal zone (Fig. 3.9).

Oviduct: This is a tubular structure connecting the ovary to the columella or uterus. It is the pathway for transporting mature eggs to the uterus for fertilization (Fig. 3.9D).

Columella: This is short tubular structure connecting the oviduct to the uterus. It is made of uterine gland cells and is called the *crustaformeria*. The cells of the *crustaformeria* may be arranged irregularly with no particular pattern, or they may be arranged in three (*tricolumella*) or four rows (*quadricolumella*).

Spermatheca: This is a swollen structure at the base of the ovary. In some groups of nematodes the spermatheca may be pouch- or sac-like. Its function is to store sperm (Fig. 3.9D).

Uterus: This is a tubular, sac-like structure connecting the spermatheca on one side and the vagina on the other. In some nematode groups such as Diplogastrina, the uterus is divided into two distinct parts: muscular and glandular. One or more developing eggs can be seen in the uterus. The number of developing eggs present in the uterus is much higher in the case of animal-parasitic species (Fig. 3.9D).

Vagina: This is a tubular structure connecting the uterus to the vulval opening: the vulva. It may be directed anteriorly or posteriorly or it may be perpendicular to the body axis. It is lined with cuticle and is surrounded by sphincter muscles (Fig. 3.9C).

Vulva: This is the opening of the female genital system to the exterior and is located midventrally. Its position in the body is of high taxonomic value. Generally it is located in the midbody region; however it may be located in the anterior half of the body or it may be situated far posterior or even further (*Meloidogyne*, *Heterodera*, *Globodera*, etc.). Generally, the vulva is circular or oval, but a transverse or longitudinal slit-like vulva is also common (Fig. 3.9D).

Male: The male reproductive system has many more components than that of females, and it is also of more taxonomic importance. In some nematode groups light microscopic studies of only females are not enough to identify up to the species level. Male sexual characteristics can broadly be divided into primary and secondary. Primary sexual characteristics are composed of the testis, seminal vesicle, ejaculatory duct, and cloacal chamber and its associated glands; secondary sexual characteristics include the spicules, gubernaculum, lateral guiding pieces, copulatory muscles, genital papillae, and bursa (Fig. 3.10).

Primary Sexual Characteristics
Testis: This is an elongated, tubular structure placed most distally from the cloaca. Its function is the production of male gametes: sperm. The testis may or may not be paired. If there is only one testis it is said to be *monorchic*, and if there are two the condition is called *diorchic*. In the monorchic condition, the testis is outstretched with the tip directed anteriorly; in the diorchic condition one of the testes is reversed with the whole of the testis directed posteriorly. The tip of the testis acts as the *germinal zone*, whereas the region posterior to it is the *growth zone*. Synthesis of spermatocytes takes place in the germinal zone, and further development resulting in the formation of spermatozoa takes place in the growth zone (Fig. 3.10).

Seminal Vesicle: This is a tubular structure of variable length just posterior to the testis. Maturation of the spermatozoa takes place in the seminal vesicle.

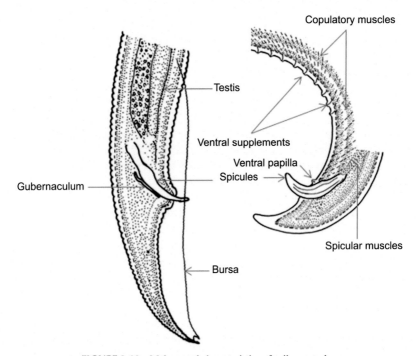

FIGURE 3.10 Male sexual characteristics of soil nematodes.

Ejaculatory Duct: This lies at the posterior end of the seminal vesicle. It is also called the *vas deferens*. The ejaculatory duct is usually associated with some secretory glands that produce an adhesive material that helps hold the female during copulation.

Cloacal Chamber: This is the place where both digestive and reproductive systems open.

Secondary Sexual Characteristics
Spicules: These are mostly paired, cuticularized, protrusible structures present in the cloacal chamber. A single spiculum may be found in some parasitic forms. The absence of spicules can also be seen among animal-parasitic forms, and in particular in the free-living genus *Myolaimus*. In most nematodes both spicules are of the same shape and size. However, in some free-living and parasitic forms the two arms of the spicules are of different shapes and sizes. The length, shape, and size of spicules vary in different species, and hence are taxonomically important (Fig. 3.10).

Gubernaculum: This is a cuticularized structure present on the posterior-dorsal side of the spicules. It may be simply trough-like or plate-like, having the form of a groove. In some free-living forms there may be a filamentous, ring-like structure known as the *distal sleeve* through which the spicules move. The gubernaculum is not necessarily present in all groups (Fig. 3.10).

Lateral Guiding Pieces: These are paired cuticularized structures present on the lateral sides of the spicules in the distal region. They are found in nematodes in which the gubernaculum is absent. They guide the movement of the spicules.

Copulatory Muscles: These are oblique bands of muscles present in the subventral, precloacal region. The more posterior curvature of the body in heat-killed males is due to the presence of copulatory muscles (Fig. 3.10).

Genital Papillae: These are minute, papillae-like projections from the body cuticle anterior or posterior to the cloaca. They may be present laterally, subventrally, ventrally, and subdorsally, but never dorsally. Except for the ventral papilla/papillae, all other papillae (lateral, subventral, and subdorsal) are paired. The ventral ones, when they are unpaired, whether modified or not, are generally called *ventromedian supplements* (e.g., Dorylaimida), and are of different shapes and sizes. Genital papillae are innervated with nerve fibers and have a sensory function. The presence or absence and the number and arrangement of the genital papillae are important taxonomic characteristics (Fig. 3.10).

Bursa: This is a paired, wing-like cuticular expansion in the region of tail. It is present in most tylenchid nematodes and in some free-living forms. Depending on the size, the bursa can be *adanal* (a small bursa present in the cloacal region only), *leptoderan* or *subterminal* (a large bursa but not covering the tail tip), and *peloderan* or *terminal* (covering the entire tail) (Fig. 3.10).

Tail: The tail is defined as a postanal or postcloacal axial elongation of the body. The cuticle in the tail region is thicker than the body cuticle. The tail has a great variety of shapes, sizes, lengths, etc. It may be short, digitate, clavate, conoid (Fig. 3.11A), hemispheroid, long conoid (Fig. 3.11B), long, whip-like, filamentous, etc. (Fig. 3.11C). Therefore, it is of great taxonomic importance. The tail may be simple or have several associated structures such as *phasmids* (paired minute, papillae-like structures present in the lateral fields) (Fig. 3.11D), *scutella* (a kind of phasmid that is bigger in size) (Fig. 3.11E), *caudal glands* (usually three unicellular glands present in the tail with a small opening at the tail tip), *caudal pores*, *caudal setae*, *spinnerets* (openings of caudal glands), *mucros* (spine-like projections from the tail tip especially with a rounded terminus), etc. Some species may show sexual dimorphism in the tail, generally with a longer and straighter tail in females.

For ecological studies using nematodes as biological models, identification up to at least the generic level is required. For any ecologist it is easy to understand the genus and species names because both are italicized or underlined.

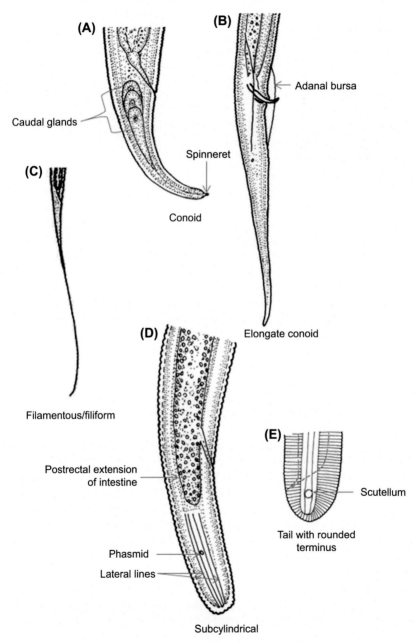

FIGURE 3.11   Different types of tails of soil nematodes: (A) conoid; (B) elongate conoid; (C) filamentous/filiform; (D) subcylindrical; (E) tail-rounded terminus.

However, in the higher categories/taxa (singular = taxon) the system is different. It is therefore necessary to understand the Pearse system of naming the higher taxa that was advocated by Chitwood (1958). The stem name/taxon, as per this system, takes different endings as follows: phylum–a; class–ea; subclass–ia; order–ida; suborder–ina; superfamily–oidea; family–idea; and subfamily–inae.

To express the morphometric characteristics of the nematode better, De Man first suggested a system (1876). There have been since additions, modifications, and amendments made by Cobb (1914), Thorne (1949), Caveness (1964), and others. These morphometric parameters are popularly known as the De Man's formula or the De Man's indices:

n = number of measured specimens.

L = total body length.

V = distance of vulva from anterior end/total body length × 100

a = total body length/greatest body diameter.

b = total body length/length of pharynx.

b′ = total body length/distance from anterior end to base of esophageal glands.

MB% = % distance from anterior to median bulb relative to length of esophagus.

c = total body length/tail length.

c′ = tail length/tail diameter at anus or cloaca.

s = stylet length/body diameter at base of stylet.

T = % length of testis relative to body length.

· P = % distance of phasmid from anus in relation to length of tail.

P^a = % distance of anterior phasmid from anterior end in relation to body length.

P_p = % distance of posterior phasmid from anterior end in relation to body length.

G^1 = % length of anterior female gonad in relation to body length.

G_2 = % length of posterior female gonad in relation to body length.

ABD = anal body diameter(s) or body diameter at anus/cloaca.

VBD = vulval body diameter or body diameter at vulval region.

DN = dorsal esophageal gland nucleus.

DO = dorsal esophageal gland orifice.

SVN = subventral esophageal gland nuclei.

3.2 GENERAL CHARACTERISTICS OF NEMATODE FAMILIES RECORDED FROM THE GRASSLAND TRANSECT

Nematodes, which are popularly known as roundworms, are the most numerous organisms on Earth. Four of every five multicellular animals on the planet are nematodes (Platt, 1994). Their distribution ranges from deep ocean trenches to the top of high mountains, encompassing all possible habitats including hot springs, ice, and low-oxygen, acid environments

(Andrássy and Zombori, 1976). Nematodes can survive a variety of extreme conditions such as desert soils (Freckman et al., 1977), Antarctic climates (Pickup and Rothery, 1991), dry soils without hosts (Womersley and Ching, 1989), dry seeds, plant debris, and dust (Barrett, 1991). Their abundance reaches many millions of individuals per square meter in soil and bottom sediments of aquatic habitats, and a handful of soil may contain more than 50 species. Along the grassland transect, altogether 32 families were identified in different ecosystems from the desert to the desert steppe, typical steppe and meadow steppe. The main characteristics of those 32 families are given subsequently.

Cephalobidae: Cephalobidae are observed in almost every soil sample. The most important characteristic is the structure of the mouth cavity, which can be derived from the rhabditid tube but the rings or rhabdia are not fused. The cheilostom is wider than the gymnostom and stegostom; the pharynx is tripartite with a strong basal bulb; the vulva is at two-thirds of the body length; the ovary is reflexed with postvulval extension of the ovary mostly with double flexures. Only one gonad is present in the female (page 70).

Panagrolaimidae: The Panagrolaimidae prefer food rich in decaying material (Bongers, 1994). The mouth of Panagrolaimidae is composed of five rhabdia, but the rhabdia gradually increase in diameter, resulting in a funnel-shaped or even rectangular mouth cavity. The lip region has no probolae; the pharynx is tripartite with a strong basal bulb; the reflexed part of the ovary behind the vulva rarely has a simple flexure; males have five to seven pairs of genital papillae (page 81).

Monhysteridae: Monhysteridae has a smooth cuticle and a relatively long tail ratio (c′ generally exceeds 5). They can be identified on the basis of the position of the amphids and vulva, the shape of the spicules, and the tail shape. The cephalic setae are long; the amphids are circular; the stoma is funnel-shaped; the pharynx is nearly cylindrical; the genital system is monoprodelphic; the postvulval uterine sac is absent; males have no genital papillae or supplements, but there is a folded cuticle in the preanal region; caudal glands and a terminal spinneret are present (page 86).

Tripylidae: The mouth cavity is closed but a little tooth in the lumen is still visible. The stoma is usually funnel shaped and walls are devoid of heavy sclerotization; the median tooth is small; the pharyngo-intestinal junction has a large valve; caudal glands are present (page 89).

Rhabdolaimidae: The cephalic setae are small and inconspicuous; the stoma is very small; the amphids mostly have slit-like apertures; the male genital supplements are papilliform or absent (page 93).

Plectidae: The mouth cavity of Plectidae is tubular with a funnel-shaped base and is easily recognized by the conical tail with a spinneret. The cuticle is annulated and lateral fields are well demarcated; the somatic seta is more prominent on the tail; the amphids are unispiral, circular, oval, or simply slit-like; the stoma narrows posteriorly; females are amphidelphic; males are diorcheic; genital supplements are sclerotized and tubular, and rarely absent; the tail is conoid, with caudal glands and a spinneret (page 95).

Tylenchidae: This family is commonly found in soil samples and often forms the dominant group in the sample. They have only one gonad and a long tapering or filiform tail, which is about five times as long as the anal diameter, or longer. The body is 0.3–1.3 mm long; lateral fields have two to six multiple incisures and are rarely absent; the stylet is generally small and is sometimes long or very long in a few species; the basal pharyngeal bulb is pyriform; the spicules are slender; the bursa is adanal or reduced or absent; the tail is elongated conoid to long filiform, and similar in both sexes; there is a phasmid-like structure present near the vulval region outside the lateral field (page 101).

Dolichodoridae: This has a cephalic region with four lobes; the stylet is long and the conus is much longer than the shaft; the pharyngeal glands do not form a lobe over the intestine; the bursa has three lobes (page 113).

Psilenchidae: Their cuticle has distinct annules; the lateral field has four incisures and the conus is much shorter than the shaft; basal knobs may be either present or absent; the tails are filiform and are similar in both sexes; the bursa is adanal; phasmids are on the tail (page 120).

Hoplolaimidae: The mouth contains a robust stylet. Distinguishing characteristics include the protruding lip area and a heavily sclerotized head; females have two gonads and a pharyngeal gland that overlaps the intestine. Sexual dimorphism is present in the anterior end; males have less developed anterior ends than do females; the stylet is rounded and indented, sometimes with anchor- or tulip-shaped knobs; pharyngeal glands usually overlap the anterior intestine; the bursa is large, rarely subterminal (page 127).

Pratylenchidae: Nematodes in this family are endoparasitic, mainly living in the roots of plants. Vermiform, mature females are rarely swollen; the cuticle has prominent annulations; the labial framework is heavily sclerotized; the median pharyngeal bulb is strongly developed, pharyngeal glands extend over the intestine (with the exception of a few species of *Pratylenchoides*). The tail is twice as long as the anal body diameter (except for *Nacobbus*) (page 134).

Anguinidae: Many species of this family are capable of parasitizing the aerial parts of plants. Mature females may be swollen. Esophageal glands are located in the basal bulb. The tail in both sexes is similar, conoid to filiform, and is exceptionally subcylindroid. The bursa (in males) is adanal to subterminal, never reaching the tail tip (page 138).

Criconematidae: Nematodes in this family are relatively easily recognized owing to their stockiness and deeply ringed body. Their body is sausage shaped or cylindrical; the cuticle is thick with prominent retrorse annules, with or without scales or spines; the stylet is massive; the conus is much longer than the shaft; basal knobs are anchor shaped or slope backward; the isthmus is very short; and the glandular pharyngeal basal bulb is greatly reduced (page 142).

Paratylenchidae: The stylet knobs are close to the metacorpus valve. The stylet conus is much longer than the shaft; the pharynx has a broad procorpus; the female is vermiform or saccate or obese; the males and juveniles are vermiform and strongly ventrally curved on fixation; the tail is short (page 145).

Aphelenchidae: The Aphelenchidae and Aphelenchoididae are easily distinguished from other stylet bearers by their strongly developed and light-refractive median bulb. The stylet has no basal knobs; the pharyngeal glands may or may not overlap the intestine; and the dorsal pharyngeal gland nucleus opens in the median bulb. Males may have a bursa or not; if it is present, it is supported by four bursal ribs (page 148).

Aphelenchoididae: Pharyngeal glands overlap the intestine dorsally; spicules are thorn shaped; and the gubernaculum is absent. The bursa is usually absent; if it is present, it is terminal or sometimes represented by bursal folds. Genital papillae are in one to three pairs (page 151).

Dorylaimidae: Almost all members of the following families (17–30) except Alaimidae are found in undisturbed soils. Dorylaims are large, stout nematodes, usually with a long, filiform tail in the female, which exhibits sexual dimorphism. The odontostyle is not attenuated, usually as long as the width of the lip region. The tail of the male is bluntly rounded (page 160).

Aporcelaimidae: The Aporcelaimidae are characterized by a thick cuticle and folded membranous guiding ring. The aperture of the odontostyle is wide, usually more than half of its length (page 163).

Qudsianematidae: These are medium-sized nematodes with a comparatively wide lip region with separate lips. The odontostyle has a wide aperture and fixed guiding ring. The tail is short and is similar in both sexes (page 171).

Nordiidae: Members of the family Nordiidae are characterized by a relatively attenuated odontostyle, usually longer than the width of the lip region. The aperture is small, usually less than 33% of the length of the odontostyle. The guiding ring is sclerotized and lips are usually compact (page 180).

Longidoridae: The family is relatively easily recognized because of its length and slenderness; the odontostyle is very long with a single guiding ring and the odontophore is not flanged basally (page 187).

Xiphinematidae: The odontostyle is very long with a double guiding ring; the odontophore is flanged basally (page 189).

Belondiridae: This family is characterized by a muscular sheath around the basal broader part of the pharynx and an axial spear. The posterior expanded part of the esophagus is covered by a thick sheath of spiral (rarely longitudinal) muscles (page 191).

Tylencholaimidae: The cuticle is loose and the lip region is cap-like. The odontostyle is small, or long or robust. The expanded part of the pharynx is about one-half or less of the pharyngeal length (page 196).

Leptonchidae: It is a fungal feeder that often occurs in arid sandy soils. The odontostyle is symmetrical, attenuated, often solid, and needle-like; the pharyngeal bulb is usually pyriform (page 197).

Mydonomidae: Members of this family are easily recognized by the presence of an asymmetrical odontostyle (the dorsal wall is longer than the ventral one), the distinct aperture of the odontostyle, and the cylindroid basal expanded part of the pharynx (page 203).

Nygolaimidae: The Nygolaimidae family is characterized by the mural tooth, which is placed on the right subventral wall of the mouth cavity. The basal expanded part of the pharynx is about one-half the pharyngeal length; cardiac glands are present (page 205).

Campydoridae: The feeding apparatus has a mural tooth located on the subdorsal wall of the buccal cavity; the basal part of the pharynx is small with a strongly developed triquetrous chamber (page 210).

Diphtherophoridae: The body is short and plump; the spear has a basal swelling; the pharynx has anterior slender and oblong or pyriform posterior parts; the spicules are triploid; the ventromedian supplements are weakly developed (page 212).

Prismatolaimidae: The cuticle has fine annulations; somatic setae are present; the lip region is continuous with the body; cephalic setae are well developed, amphidial apertures are slit-like, the stoma is wide and cylindrical, and narrows posteriorly. The pharynx is muscular and tubular, gradually expanding posteriorly to join the pharyngo-intestinal junction (page 215).

Mylonchulidae: The buccal cavity is wide and spacious; the dorsal tooth is in the anterior half or middle of the buccal cavity; one or several rows of denticles are arranged transversely on the subventral walls; the tail is short and conoid, with or without a spinneret (page 219).

Alaimidae: The cuticle is smooth or, rarely, has longitudinal ridges; the lip region is continuous with the body; the stoma is vestigial without an armature; there are a few to many ventromedian supplements; there is a single outstretched testis (page 222).

3.3 NEMATODE GENERA AND SPECIES PRESENT IN THE GRASSLAND TRANSECT

Altogether 66 nematode genera were identified along the grassland transect. They belong to Rhabditida, Monhysterida, Enoplida, Araeolaimida, Tylenchida, Aphelenchida, Dorylaimida, Triplonchida, Monochida, and Alaimida. Detailed taxonomic information and illustrations of each genus or species are provided in this section.

Order Rhabditida Chitwood, 1933

Diagnosis: The cuticle is usually annulated, rarely ornamented with longitudinal striae or punctations. The lip region is continuous; the lips are separate, three or six, often with projections. Amphids are small and located on lateral lips. Stomas are mainly of two types: tuboid or more or less spacious; in the former case they are unarmed or possess minute denticles; in the latter case they are usually provided with well-developed teeth. The pharynx possesses a median or terminal valvular bulb. The excretory pore is visible. The intestine has a wide lumen. Three rectal glands are generally present. The female genital system is amphidelphic or monoprodelphic; a postuterine sac is generally present in the latter case. The ovaries are reflexed and oviparous or viviparous. The spicules are occasionally fused. Males either have paired genital papillae or a caudal bursa possessing paired rodlike papillae or ribs. The tail often shows sexual dimorphism, but with distinct phasmids. Caudal glands or spinnerets are absent.

Type suborder

Rhabditina Chitwood, 1933*

Other suborders

Cephalobina Andrássy, 1974*
Diplogastrina Micoletzky, 1922

Myolaimina Inglis, 1983
Teratocephalina Andrássy, 1974*

Indicates the suborder, superfamily, and family with representative genera in our collection.

Suborder Cephalobina Andrássy, 1974

Diagnosis: Three or six lips, mostly separate; projections are present on the labial region (probolae) of various appearances. Amphids are small and pore-like on lateral lips. The stoma is tuboid and is composed of six rings: cheilo-, gymno-, pro-, meso-, meta-, and telostom. The stegostom is surrounded by a pharyngeal collar. The dorsal wall of the metastom has a minute tooth-like projection. The corpus and isthmus are well separated, with a well-developed grinder present in the terminal bulb. The excretory pore is distinct. The female genital system is always unpaired and prevulval, but the ovary is reflexed beyond the vulva. The spermatheca is generally present at the anterior flexure of the gonad and is predominantly oviparous. Spicules are simple, never fused. The gubernaculum is present. Male supplements are papilloid, arranged in pairs. The bursa is absent. The tail is generally short. Phasmids are well discernible.

Type superfamily

Cephaloboidea Filipjev, 1934*

Other superfamilies

Chambersielloidea Thorne, 1937
Panagrolaimoidea Thorne, 1937*

Superfamily Cephaloboidea Filipjev, 1934

Diagnosis: The lip region is of the simple amalgamated type or has an elaborate modified structure. The stoma is narrow with uniform sclerotized ring elements divided into three distinct sections, with or without a metastomal tooth. The pharynx may or may not have a grinder apparatus in the basal bulb. The female genital system is monoprodelphic, with the ovary reflexed extending beyond the vulva; in most cases the postvulval part shows double flexures; the spermatheca is invariably present at the anterior flexure of the gonad. The testis is single with a reflexed terminal part. The spicules are ventrally curved with velum and capitulum. A gubernaculum is present. There is no bursa. Genital papillae are either present or absent.

Type family

Cephalobidae Filipjev, 1934*

Other families

Elaphonematidae Heyns, 1962
Osstellidae Heyns, 1962

Family Cephalobidae Filipjev, 1934

Diagnosis: The cuticle is annulated; lateral fields are clearly demarcated, occasionally divided by longitudinal lines. There are three cephalic probolae and three or six labial probolae. The stoma is tubular and generally narrow. The cheilostom is wider than the gymnostom and stegostom. A minute tooth is present on the dorsal wall of the stegostom. The pharynx usually consists of three sections bearing a strong bulb. The vulva is located at two-thirds of the body length, with the ovary reflexed, extending far posterior to the vulva; the postvulval extension of the ovary mostly has double flexures. A postvulval uterine sac is present and generally short. Males are almost as frequent as females. Phasmids are distinct.

Type genus

Cephalobus Bastian, 1865

Key characteristics of some commonly found genera of Cephalobidae

1. Six sharply pointed lips equal in size; tail is pointed .... *Eucephalobus*
2. Three blunt lips are asymmetrical and unequal in size; the tail is bluntly rounded, sometimes with a mucro..................... *Cephalobus*
3. Dorsal labial probolae bifurcate, pharynx with almost cylindrical corpus .... *Chiloplacus*
4. The lip region has low, rounded, or conical, undivided labial probolae; the pharynx has an elongated corpus.................... *Acrobeloides*
5. The lip region has six petal-like probolae with a triangular tip and a pedunculate base; guiding processes are present; a ventral guarding process is prominent......................... *Acromoldavicus*
6. The cephalic probolae are pointed and flap-like; the labial probolae are not swollen at the base .... *Acrobeles*
7. The cephalic probolae have serrated and cuticularized margins; the labial probolae are divided near the middle.................. *Cervidellus*
8. The cephalic probolae do not have serrated margins; the labial probolae are fork-like .... *Acrobelophis*
9. The tip of the labial probolae is arcuate and fork-like; the tail of the female is broadly rounded .... *Stegelleta*
10. The cephalic probolae is flap-like and deeply notched with a dentate projection at the base; the postvulval uterine sac is absent...... *Zeldia*

Genera recorded

Acrobeles Linstow, 1877
Acrobeloides (Cobb, 1924) Thorne, 1937
Acromoldavicus Nesterov, 1970
Cephalobus Bastian, 1865
Cervidellus Thorne, 1937
Chiloplacus Thorne, 1937
Eucephalobus Steiner, 1936

Genus *Acrobeles* Linstow, 1877 (Figs. 3.12 and 3.13)

Diagnosis: The body is small to large (L=0.3–1.1 mm). The cuticle is single or double, with large annules, with or without longitudinal striae, punctations, and/or pores. The lateral field has two or three incisures; if the cuticle is double, often it has undulating internal pseudolines. Amphids are relatively distinct and circular. The labial probolae are long and deeply bifurcated. Each prong has at least seven tines; each tip usually has two elongate, separated apical tines. The cephalic probolae are high, triangular, separate, and fringed by numerous tines. The stoma cephaloboid has distinct cheilorhabdia that is large and spherical upon cross-section. The pharyngeal corpus is cylindrical to fusiform. The excretory pore position varies from very anterior to the opposite basal bulb. The female genital system is cephaloboid; the spermatheca and postuterine sac are small to large. The vulva is flush with the body, occasionally sunken. Males have three pairs of precloacal papillae, five pairs of postcloacal papillae, and one median papillae on the precloacal lip. Tails in both sexes are conical, usually with an acute tip. This genus is prevalent in arid soil.

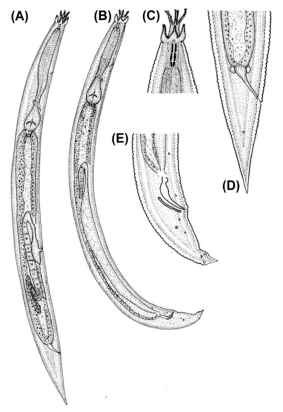

FIGURE 3.12 *Acrobeles* sp.: (A) entire female; (B) entire male; (C) anterior end; (D) female posterior end; (E) male posterior end.

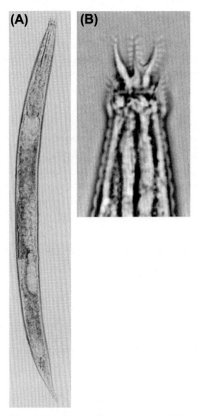

FIGURE 3.13 *Acrobeles* sp.: (A) entire female; (B) anterior end.

Type species

Acrobeles ciliatus Linstow, 1877

Commonly found species of *Acrobeles* Linstow, 1877

A. *complexus* Thorne, 1925
A. *ciliatus* Linstow, 1877
A. *elaborates* Thorne, 1925

Genus *Acrobeloides* (Cobb, 1924) Thorne, 1937
(Figs. 3.14 and 3.15)

Diagnosis: The body length varies from 0.3 to 1.2 mm. The cuticle is annulated; lateral fields each have two to five incisures extending generally to the tip of the tail. There are three lips; the labial probolae is hemispheroid or conoid, with the point always unitipped. The cephalic probolae are present but low, not strongly differentiated. The proximal half of the esophagus is fusiform. The stoma is

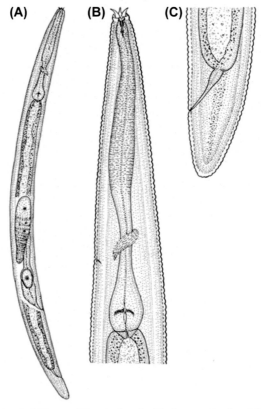

FIGURE 3.14 *Acrobeloides* sp.: (A) entire female; (B) pharyngeal region; (C) posterior region.

cephaloboid and narrow. The vulva is nearly two-thirds of the body length; the ovary has double postvulval flexures. Males are mostly unknown. The tail is short and plump, broadly rounded or conoid.

Type species

Acrobeloides buetschlii (De Man, 1884) Steiner and Buhrer, 1933

Commonly found species of *Acrobeloides* (Cobb, 1924) Thorne, 1937

A. tricornis Thorne, 1925 *nanus* (De Man, 1880) Anderson, 1968

Genus *Acromoldavicus* Nesterov, 1970 (Figs. 3.16 and 3.17)

Diagnosis: The body length is usually less than 1 mm and is straight or slightly ventrally curved. The cuticle has coarse transverse annulations and fine longitudinal lines. The lateral fields each have three incisures. Amphids have inconspicuous, slit-like openings. The lips are well-developed, modified into

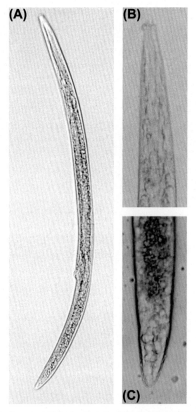

FIGURE 3.15 *Acrobeloides* sp.: (A) entire female; (B) pharyngeal region; (C) posterior region.

flap-like structures. The lip region is slightly set off from the body. The stoma is short, simple, and tubular. The anterior part of the pharynx is muscular; the isthmus is moderately long, expanding posteriorly to a valvate basal bulb. The vulva is postmedian. The female genital system is monoprodelphic; the ovary is reflexed, extending well beyond the vulva. A postvulval uterine sac is present. Males have a single testis, arcuate spicules, a simple gubernaculum, and eight pairs of genital papillae. Sexual dimorphism in the tail is present; females have a pointed tail terminus and males have a rounded tail terminus.

Type species

Acromoldavicus skrjabini (Nesterov and Lisetskaya, 1965) Nesterov, 1970

Remarks: So far, only two species, viz., *Acromoldavicus mojavicus* Baldwin, De Ley, Mundo-Ocampo, De Ley, Nadler and Gebre, 2001 and *A. skrjabini* (Nesterov and Lisetskaya, 1965) Nesterov, 1970, are recognized. This is the first report of the occurrence of this small and rare genus in China.

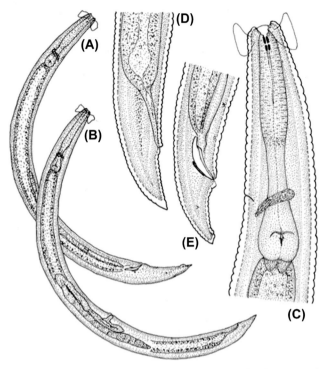

FIGURE 3.16 *Acromoldavicus* sp.: (A) entire male; (B) entire female; (C) pharyngeal region; (D) female posterior end; (E) male posterior end.

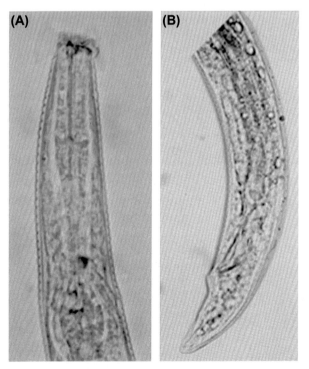

FIGURE 3.17 *Acromoldavicus* sp.: (A) pharyngeal region; (B) male posterior end.

Genus *Cephalobus* Bastian, 1865 (Figs. 3.18 and 3.19)

Diagnosis: The body tapers slightly at the extremities. The cuticle has prominent transverse striations. Lateral lines extend beyond the phasmids or reach the tail tip. The lip region is somewhat lobed without labial probolae. There are six lips that are unequal in size. The stoma is cephaloboid. The pharynx has a long, muscular, cylindrical corpus, an isthmus, and a basal bulb containing a simple valvular apparatus. The vulva is located in the posterior third of the body. The female genital system is cephaloboid. Spicules are slightly arcuate and somewhat fusiform. Females have a uniformly tapering tail with blunt terminus. The male tail is more sharply tapered and more ventrally curved. Genital papillae have six pairs.

Type species

Cephalobus persegnis Bastian, 1865

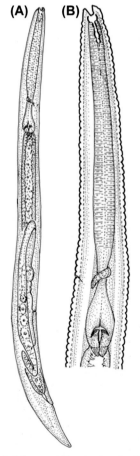

FIGURE 3.18 *Cephalobus* sp.: (A) entire female; (B) pharyngeal region.

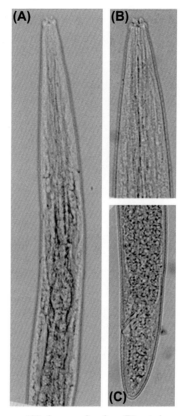

FIGURE 3.19 *Cephalobus* sp.: (A) pharyngeal region; (B) anterior end; (C) female posterior end.

Commonly found species of *Cephalobus* Bastian, 1865

C. persegnis Bastian, 1865
C. parvus Thorne, 1937

Genus *Cervidellus* Thorne, 1937 (Figs. 3.20 and 3.21)

Diagnosis: Very small nematodes range from 0.3 to 0.5 mm. The cuticle is finely annulated; lateral fields each have two, three, or five incisures. Lip margins have U-shaped refractive elements; there are six cephalic probolae that are triangular or leaf-like. Labial probolae are thin and Y shaped, occasionally with secondary tines. The pharyngeal corpus is cylindrical. The female genital system is cephaloboid. The tails of both sexes are conoid with a pointed tip.

Type species

Cervidellus cervus (Thorne, 1925) Thorne, 1937

Commonly found species of *Cervidellus* Thorne, 1937

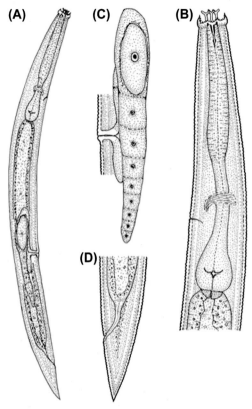

FIGURE 3.20   *Cervidellus* sp.: (A) entire female; (B) pharyngeal region; (C) female genital system; (D) posterior region.

C. vexilliger (De Man, 1880) Thorne, 1937
C. hamatus Thorne, 1937

Genus *Chiloplacus* Thorne, 1937 (Figs. 3.22 and 3.23)

Diagnosis: The body is 0.3–1.0 mm long. The cuticle is annulated; the lateral fields each have three to six lines. There are three lips; the labial probolae are biacute or bifurcate (apically incised), symmetrical, or asymmetrical, with broad "shafts." The cephalic probolae are small, generally with two incisures. The pharyngeal corpus is cylindrical. The female genital system is cephaloboid; the postvulval uterine sac is variable in length. The female tail is straight, short, and cylindroid, and the terminus is broadly rounded. The male tail is ventrally bent. Phasmids are posterior to the lateral caudal papilla.

Type species

Chiloplacus symmetricus (Thorne, 1925) Thorne, 1937

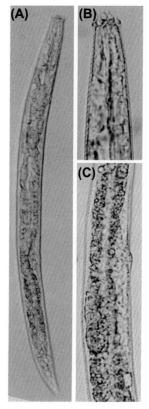

FIGURE 3.21 *Cervidellus* sp.: (A) entire female; (B) pharyngeal region; (C) vulval region.

Commonly found species of *Chiloplacus*

C. symmetricus (Thorne, 1925) Thorne, 1937
C. demani (Thorne, 1925) Thorne, 1937
C. trilineatus Steiner, 1940

Genus *Eucephalobus* Steiner, 1936 (Figs. 3.24 and 3.25)

Diagnosis: The body tapers slightly at the extremities. The cuticle has prominent transverse striations. The lateral lines do not extend beyond the phasmids. The lip region is somewhat lobed without labial probolae. There are six lips that are symmetrical in their size. The stoma is cephaloboid. The pharynx has a long, muscular, cylindrical, corpus; the isthmus and basal bulb contain a simple valvular apparatus. The vulva is located in the posterior third of the body. The female genital system is cephaloboid. The spicules are slightly arcuate and somewhat fusiform. Females have a uniformly tapered tail with a pointed or, rarely, blunt terminus. The male tail is sharply tapered and more ventrally curved.

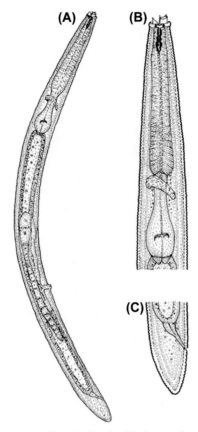

FIGURE 3.22   *Chiloplacus* sp.: (A) entire female; (B) pharyngeal region; (C) posterior region.

Type species

Eucephalobus oxyuroides (De Man, 1876) Steiner, 1936

Commonly found species of *Eucephalobus*

E. mucronatus (Kozlowska and Roguska-Wasilewska, 1963) Andrassy, 1967
E. oxyuroides (De Man, 1876) Steiner, 1936
E. striatus (Bastian, 1865) Thorne, 1937

Superfamily Panagrolaimoidea Thorne, 1937

Diagnosis: The labial margins are smooth and rarely lobed. The lips are simple without probolae; the labial papillae are minute. The anterior part of the stoma is wider than the posterior one. The cheilostom is not cuticular- ized. The gymnostom is heavily cuticularized. The female genital system is monoprodelphic. The postvulval part of the ovary is straight and rarely has

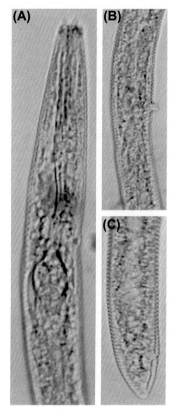

FIGURE 3.23 *Chiloplacus* sp.: (A) pharyngeal region; (B) vulval region; (C) posterior region.

flexures. Phasmids are usually located in the posterior one-third of the tail. The bursa is absent.

Type family

Panagrolaimidae Thorne, 1937*

Other families

Alirhabditidae Suryawanshi, 1971
Brevibuccidae Paramonov, 1956

Family Panagrolaimidae Thorne, 1937

Diagnosis: The cuticle is annulated; the lateral fields are distinct. The lip region does not have probolae; there are three or six lips. There are lateral amphids. The stoma consists of spacious cheilostom and gymnostom; the stegostom tapers with a small tooth-like projection. The pharynx has a corpus, isthmus, and basal bulb. The excretory pore is conspicuous. The female genital system

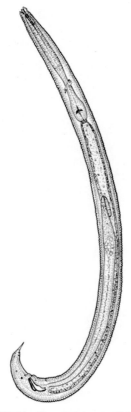

FIGURE 3.24 *Eucephalobus* sp.: Entire male.

is monoprodelphic; the reflexed part extends behind the vulva, rarely with a simple flexure. Males are mostly abundant. Preanal genital papillae have five to seven pairs. The tail is conoid to elongate, generally shorter in males. Phasmids are always distinct.

Type genus

Panagrolaimus Fuchs, 1930

Key characters of some commonly found genera of Panagrolaimidae

1. The lip region does not have deep grooves; the edges of the lip region are not cuticularized; the area of the metacorpus is not swollen; the promesostom is fused, forming a chamber . . . . *Panagrolaimus*
2. The lip region does not have deep grooves; the area of the metacorpus is swollen; the prostom and mesostom are separate with distinct rhabdions . . . . *Procephalobus*
3. The lip region has deep grooves; the edges are cuticularized; the tail is 7–8 ABDs long; the postvulval uterine sac is absent . . . . *Panagrobelium*

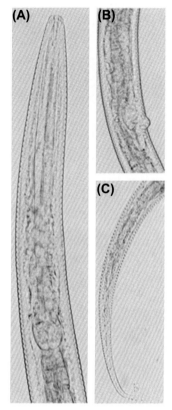

FIGURE 3.25 *Eucephalobus* sp.: (A) pharyngeal region; (B) vulval region; (C) female posterior region.

4. The lip region has deep grooves; the edges are cuticularized; the tail is 3 ABDs long; a postvulval uterine sac is present . . . . *Panagrobelus*
5. The promesostom is short; several denticles are on the metarhabdion; spicules have bifid tips; the anterior knobbed end is usually hooked ventrally; a postvulvar sac is present . . . .*Panagrellus*

Genus recorded

Panagrellus Thorne, 1938

Genus *Panagrellus* Thorne, 1938 (Figs. 3.26 and 3.27)

Diagnosis: Lips are fused or very slightly separated; labial papillae are small. The stoma is short and cylindrical. Several minute denticles are present on the stegostomal wall. The stegostom narrows posteriorly. The pharynx has a long, broad, muscular, anterior part; a short isthmus; and an ovoid basal bulb housing strong valve plates. The female genital system is monoprodelphic; the ovary is reflexed.

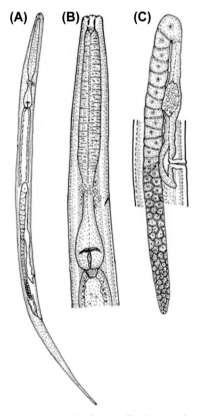

FIGURE 3.26 *Panagrellus* sp.: (A) entire female; (B) pharyngeal region; (C) vulval region.

The tip of the ovary almost reaches the rectum and the postvulval uterine sac is short. The female tail is uniformly conical with a finely pointed terminus. Males have large, cephalated, slightly arcuate, distally bifid spicules and a simple gubernaculum. The tail is divided into two parts: an anterior conical part and a distal whip-like part with a pointed terminus. Genital papillae have five to seven pairs.

Type species

Panagrellus pycnus Thorne, 1939

Commonly found species of *Panagrellus* Thorne, 1939

P. redivivus (Linné, 1767) Goodey, 1945

Order Monhysterida De Coninck and Schuurmans Stekhoven, 1933

Diagnosis: The cuticle is smooth or annulated; annules have no punctations or dots. The cephalic setae are arranged into a 6 + 4 pattern, rarely more. Amphids are circular and sometimes spiral. The stoma is small, funnel shaped, or

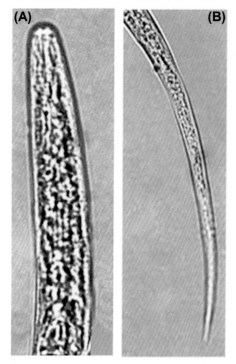

FIGURE 3.27   *Panagrellus* sp.: (A) pharyngeal region; (B) female posterior region.

spacious. Teeth or denticles are rarely present. The pharynx is cylindrical, tube-like, or swollen at the base, but a true bulb is never found. The female genital system is predominantly monoprodelphic and rarely amphidelphic, with ovary or ovaries outstretched. Male genital supplements are simple or nonexistent. There are three caudal glands with a terminal spinneret.

Type suborder

Monhysterina De Coninck and Schuurmans Stekhoven, 1933*

Other suborder

Linhomoeina Andrássy, 1974

Suborder Monhysterina De Coninck and Schuurmans Stekhoven, 1933

Diagnosis: The cuticle is smooth or annulated; annules never resolve into punctations or dots. The cephalic setae are in a 6+4 basic arrangement, rarely more. Amphids are mostly circular. The stoma is small, often spacious, and funnel shaped. Teeth or denticles are rarely found. The pharynx has no bulb-like swelling. The female genital system is monoprodelphic; the ovary is straight. The tail has a sclerotized terminal spinneret.

Type superfamily

Monhysteroidea De Man, 1876*

Other superfamily

Sphaerolaimoidea Filipjev, 1918

Superfamily Monhysteroidea De Man, 1876

Diagnosis: The cuticle is smooth or annulated without punctations or dots. Six postlabial setae are always longer than four submedial setae. Amphids are mostly circular. The stoma is small and often spacious and funnel shaped. Teeth or denticles are rarely found. The pharynx has no bulb-like swelling. The female genital system is monoprodelphic; the ovary is straight. The tail has a terminal spinneret.

Type family

Monhysteridae De Man, 1876*

Other family

Xyalidae Chitwood, 1951

Family Monhysteridae De Man, 1876

Diagnosis: Body length is between 0.3 and 1.5 mm. The cuticle has fine scattered setae that are thin and smooth. There are 10 cephalic setae: six long and four short ones; longer setae are often articulated. Amphids are circular. Ocelli are often present in limnic species, paired or unpaired. The stoma is small and funnel shaped, without armature; it rarely has minute denticles and is hardly sclerotized. The pharynx muscular, which is nearly cylindrical, occasionally has a terminal bulb-like swelling. The female genital system is monoprodelphic and is situated on the right side of the intestine, with the ovary outstretched. The postvulval uterine sac is absent. There are fewer males than females, or they are even absent in several species. Male gonads are monorchic. Spicules vary in length and are occasionally very long and thin. The gubernaculum is short. Genital papillae or supplements are absent, but the preanal region of the male has a folded cuticle. Tail in both sexes are similar: long and often filiform, with three caudal glands and a terminal spinneret. Terminal setae are either present or absent.

Type genus

Monhystera Bastian, 1865

Key characteristics of some commonly found genera of Monhysteridae

1. The pharynx has terminal swelling; the vulva is at the midbody length; the gonad is short; often the spinneret is elongated. . . . . *Monhysterella*
   The pharynx has no swollen terminal bulb; the vulva is posterior; the gonad is long; the spinneret is short. 2

2. The vulva is nearly 75% from the anterior end; the gonad is long; the rectum is heavily muscular, 2–3 ABDs long; the tail tip is not swollen.. *Geomonhystera*
   The vulva is 69–72% from the anterior end; the rectum is thin without strong musculature, usually 1 ABD long; the tail tip is swollen........3
3. Spicules are more than 2.3 ABDs long; amphids are generally at less than one lip region; the body is filled with crystals; the tail is shorter than one vulva–anus distance .... *Monhystera*
   Spicules are shorter than 2.2 ABDs; somatic setae are scattered; amphids are located at least one lip region from the anterior end; the tail is longer than one vulva–anus distance........................ *Eumonhystera*

Genus recorded

Geomonhystera Andrássy, 1981

Genus *Geomonhystera* Andrássy, 1981 (Figs. 3.28 and 3.29)

Diagnosis: The body length is between 0.5 and 1.2 mm. The cuticle is smooth or very finely annulated, provided with scattered setae. The body cavity has no crystalloids. The labial region is as wide as or sometimes wider than the adjacent body; lips are completely amalgamated; labial papillae are setose. Longer cephalic setae are articulated, half to three-fourths the length of the labial width; shorter setae are simple. Amphids are large, mostly more than one head diameter from the anterior end. Ocelli are absent. The stoma is simple, with or without denticles. The pharynx has no terminal swelling. The rectum is unusually strong and muscular, more than one anal body diameter long. The vulva is at 75%–85% of body length, close to the anus, at a distance of 1–3 ABDs from the anus. The gonad never reaches the pharynx. The nematode is oviparous. Males are rare; spicules are more than one and a half times longer than the ABD; the gubernaculum is small and simple. The cuticle on the precloacal region of the male is wrinkled. The tail is rather stout, never filiform, and ventrally curved; in females it is two to three times as long as the vulva–anus distance. The spinneret is short and conoid.

Type species

Geomonhystera villosa (Butschli, 1873) Andrássy, 1981

Commonly found species of *Geomonhystera* Andrássy, 1981

G. villosa (Butschli, 1873) Andrássy, 1981

Order Enoplida (Baird, 1853) Chitwood, 1933

Diagnosis: The lip region is usually provided with setae or is sometimes absent. Amphids are generally pocket shaped. The pharynx is cylindrical or conoid or multibulbar with three or more glands. The pharyngeal glands are open in the stomal region. The females have an amphidelphic genital system; ovaries are

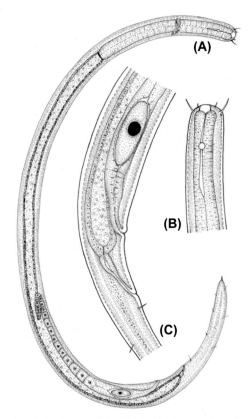

FIGURE 3.28 *Geomonhystera* sp.: (A) entire female; (B) anterior end; (C) vulval region.

reflexed. Spicules are paired or single, or even absent. Caudal glands open into the exterior through a terminal duct that is usually present. Hypodermal glands are generally well developed.

Type superfamily

Enoploidea (Baird, 1853) Sch. Stek. and De Coninck, 1933

Other superfamily

Tripyloidea (Örley, 1880) Chitwood, 1937*

Superfamily Tripyloidea (Örley, 1880) Chitwood, 1937

Diagnosis: The cuticle in the cephalic region is not double. The orifices of the dorsal pharyngeal gland and at least two subventral glands are located anterior to the nerve ring. Males have one or more papilloid supplements usually.

Type family

Tripylidae Örley, 1880*

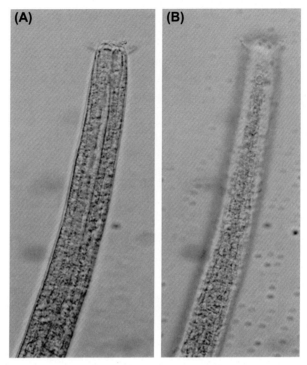

FIGURE 3.29 *Geomonhystera* sp.: (A) anterior end; (B) anterior end with amphid.

Other family

Ironidae (Filipjev, 1918) Baylis and Daubney, 1926

Family Tripylidae Örley, 1880

Diagnosis: Orifices of the dorsal pharyngeal gland and at least two subventral glands are located in the region of the stoma. The stoma is usually funnel shaped or a narrow tube with a small median tooth. The stomal walls are devoid of heavy sclerotization. The pharyngo-intestinal junction has a large valve. Males have almost straight spicules, a gubernaculum, and two or more supplements. Caudal glands are present.

Type genus

Tripyla Bastian, 1865

Key characteristics of commonly found genera of Tripylidae

1. A lip region with three flat lips; amphids are inconspicuous; the stoma is cylindrical . . . . *Tripyla*
2. A lip region with six round to conical lips; amphids are cup shaped; the stoma is funnel shaped . . . . *Tobrilus*

3. The mouth cavity has two teeth and a wide lumen . . . . *Paratripyla*

Genus recorded

Only the genus type was found in our collection

Genus *Tripyla* Bastian, 1865 (Figs. 3.30 and 3.31)

Diagnosis: The cuticle is smooth or has fine transverse striations. The lip region is continuous with the body. There are three lips; the labial papillae are short or long and bristle-like. Amphids are usually inconspicuous. The stoma is small and tubular with a simple median tooth on the dorsal wall. The pharynx is an almost uniform cylinder. The pharyngo-intestinal junction has well-developed cells. The vulva is located medially; the genital system is amphidelphic; the ovaries are reflexed. The males have a pair of testes, spicules, a gubernaculum, and a series of midventral genital papillae. Caudal glands have an associated terminal duct present. The tail in both sexes is conoid and ventrally curved with a blunt terminus.

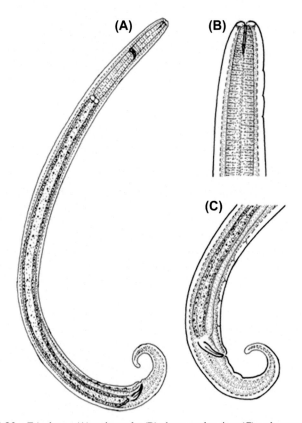

FIGURE 3.30 *Tripyla* sp.: (A) entire male; (B) pharyngeal region; (C) male posterior region.

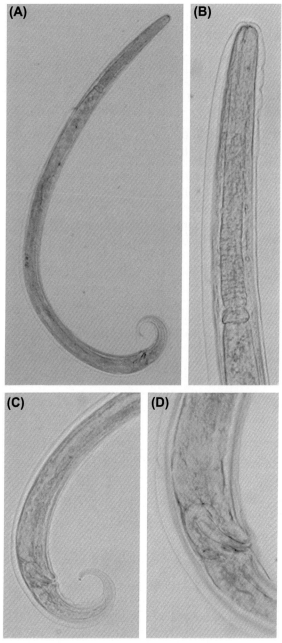

FIGURE 3.31 *Tripyla* sp.: (A) entire male; (B) pharyngeal region; (C) male posterior region; (D) cloacal region.

Type species

Tripyla glomerans Bastian, 1865

Commonly found species of *Tripyla* Bastian, 1865

T. affinis De Man, 1880
T. filicaudata De Man, 1880

Order Araeolaimida De Coninck and Schuurmans Stekhoven, 1933

Diagnosis: The body is with or without submedial setae; the cuticle has annulations. The cephalic setae generally have four, sometimes six plus four. Amphidial apertures are open-circular, spiral, circular, or occasionally slit-like. The stoma is small, short, or elongate, and mostly without armature. The pharyngeal terminal bulb is present or absent. The female generally has amphidelphic gonads; it is rarely monodelphic, with a reflexed ovary or ovaries. Males generally have large, tubelike, protrusible, rarely papilliform supplements. The tail generally has three caudal glands, usually with and rarely without a sclerotized terminal spinneret.

Type suborder

Araeolaimina De Coninck and Schuurmans Stekhoven, 1933

Other suborder

Leptolaimina Lorenzen, 1979*

Suborder Leptolaimina Lorenzen, 1979

Diagnosis: The cephalic setae have three whorls. The stoma is simple, more or less funnel shaped. Amphids have circular or spiral apertures. The anterior pharynx has tuboid endings in pharyngeal radii; a basal pharyngeal bulb is present, usually with valve plates.

Type superfamily

Leptolaimoidea Örley, 1880

Other superfamilies

Haliplectoidea Chitwood, 1951*
Plectoidea Örley, 1880*
Metateratocephaloidea Eroshenko, 1973

Superfamily Haliplectoidea Chitwood, 1951

Diagnosis: The stoma is usually very small and provided with small teeth. The pharynx is muscular and a basal bulb is present. Male genital supplements are papilliform or absent.

Type family

Rhabdolaimidae Chitwood, 1951*

Other family

Aulolaimidae Jairajpuri and Hooper, 1968

Family Rhabdolaimidae Chitwood, 1951

Diagnosis: The cuticle is smooth or has fine annulations. The lip region is continuous with the body; the cephalic setae may be present or absent. The amphidial apertures are small, slit-like, or circular. The stoma is usually very narrow and tubular, with three small teeth. The pharynx usually has a valvate-basal bulb. Females have an amphidelphic or monoprodelphic genital system. Genital supplements in males are papilliform or absent. The tail has caudal glands and a spinneret.

Type genus

Rhabdolaimus De Man, 1880

Key characteristics of commonly found genera of Rhabdolaimidae

1. The female genital system is monoprodelphic . . . . *Udonchus*
   The female genital system is amphidelphic. .2
2. The pharynx has weak, basal, bulb-like swelling; cephalic setae are present; ovaries are outstretched . . . . *Rogerus*
3. The pharynx has a well-developed basal bulb; cephalic setae are absent; ovaries are reflexed. . . . . *Rhabdolaimus*

Genus recorded

Udonchus Cobb, 1913

Genus *Udonchus* Cobb, 1913 Syn. Monochromadora Schneider, 1937 (Figs. 3.32 and 3.33)

Diagnosis: These small-sized nematodes are usually 0.5–0.7 mm in length. The cuticle is smooth. The lip region is continuous with the body contour; the lips are fused. Amphids have oval apertures. The stoma is narrow and tubular, and is provided with a dorsal tooth and a pair of subventral teeth. The pharynx is muscular and cylindrical, and terminates in a valvate-basal bulb. The female genital system is monoprodelphic. The ovary is reflexed; the postvulval uterine sac is absent. The tail has caudal glands and a spinneret gradually tapering to an elongate, conoid terminus.

Type species

Udonchus tenuicaudatus Cobb, 1913

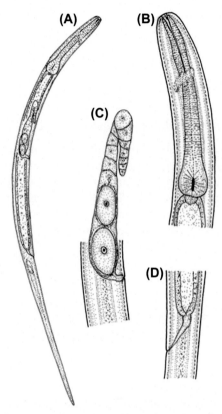

FIGURE 3.32 *Udonchus* sp.: (A) entire female; (B) pharyngeal region; (C) vulval region; (D) anal region.

Commonly found species of *Udonchus*

U. tenuicaudatus Cobb, 1913

Superfamily Plectoidea Örley, 1880

Diagnosis: The cuticle is annulated. There are four cephalic setae that are short and rarely absent. Amphidial apertures are oval, slit-like, or spiral. The stoma is tubular and unarmed. The pharyngeal terminal bulb has a strongly developed valvular apparatus. The excretory gland is in the surrounding isthmus. Male supplements when present are tubular.

Type family

Plectidae Örley, 1880*

Other family

Chronogastridae Gagarin, 1975

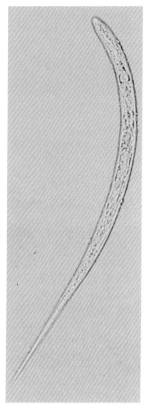

FIGURE 3.33 *Udonchus* sp.: Entire female.

Family Plectidae Örley, 1880

Diagnosis: The cuticle is annulated. Lateral fields are well demarcated. Somatic setae are generally present, more prominently seen on the tail. There are four cephalic setae of moderate length. Amphidial apertures are unispiral, circular, oval, or simply slit-like. The stoma has anterior wide and posterior narrow parts. The pharynx is muscular, with a strong valvate-terminal bulb. The cardia is well developed. The female genital system is amphidelphic. Males have diorcheic testes, arcuate spicules, and a gubernaculum. Genital supplements are generally sclerotized and tubular, and rarely absent. The tail is usually elongate conoid, with caudal glands and a spinneret.

Type genus

Plectus Bastian, 1865

Key characteristics of commonly found genera of Plectidae

1. The cephalic region is slightly set off from or continuous with the body contour; labial papillae are not setae form; the protostome is long, at least as long as it is wide, or even longer . . . .*Plectus*

2. The cephalic region is strongly set off from the body; labial papillae are setae form; the protostome is not as long as it is wide .... *Cheiloplectus*
3. The cephalic region is slightly set off from or continuous with the body contour; the cephalic setae are fine and short; the stoma has anterior globular and posterior tubular parts .... *Anaplectus*
4. Cervical expansion is a smooth, cornua lamelliform with four tines each .... *Wilsonema*

Genera recorded

Plectus Bastian, 1865
Wilsonema Cobb, 1913

Genus *Plectus* Bastian, 1865 (Figs. 3.34 and 3.35)

Diagnosis: Body size is small to medium (L=0.3–2.0 mm) with sparsely distributed somatic setae. The cuticle has clear annulations; lateral fields are well demarcated. The lip region is continuous with or set off from the body; lips are separate, with 6-minute labial papillae and four cephalic setae. Amphidial apertures are spiral or circular. The stoma is funnel shaped, usually narrowing posteriorly. The pharynx is muscular; the basal bulb is well developed. Females have an amphidelphic genital system; ovaries are reflexed. Males are usually rare or absent. Spicules are usually dissimilar. Precloacal genital supplements are generally tubular, sclerotized, or reduced, and rarely absent. The tail is conoid, with caudal glands and a spinneret.

Type species

Plectus parietinus Bastian, 1865

Commonly found species of *Plectus* Bastian, 1865

P. elongatus Maggenti, 1961
P. exinocaudatus Truskova, 1976
P. geophilus De Man, 1880
P. longicaudatus Bütschli, 1873
P. parvus Bastian, 1865
P. pusteri Fuchs, 1930
P. rhizophilus De Man, 1880
P. tenuis Bastian, 1865
P. velox Bastian, 1865

Wilsonema Cobb, 1913 (Figs. 3.36 and 3.37)

Diagnosis: These are usually very small nematodes (L=0.2–0.4 mm). Cuticular annulations are present. The body setae are short and scattered. Cervical expansion is wide. Cornua lamelliform have four tines each. The stoma is cylindrical. The pharynx is muscular and the basal bulb is well developed. The female

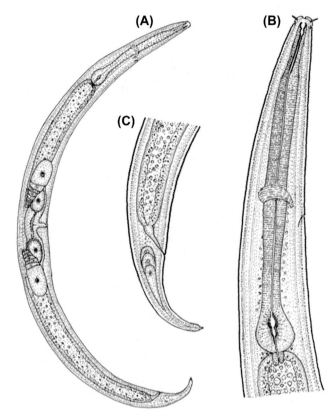

FIGURE 3.34   *Plectus* sp.: (A) entire male; (B) pharyngeal region; (C) female posterior region.

reproductive system is amphidelphic and ovaries are reflexed. Males (usually rare) have no gubernaculum or midventral supplements. Tails in both sexes are conoid, with caudal glands and a spinneret.

Type species

Wilsonema otophorum (De Man, 1880) Cobb, 1913

Commonly found species of *Wilsonema otophorum* (De Man, 1880) Cobb, 1913

W. otophorum (De Man, 1880) Cobb, 1913
W. bangalorense (Chawla, Khan and Prasad, 1969) Zell, 1993

Order Tylenchida Thorne, 1949

Diagnosis: Subclass Tylenchia. These are Secernentean nematodes with a spear, either free-living fungal and algal feeders or parasites of aerial or subterranean parts of plants and arthropod hemocoel and celomic tissues of both annelids

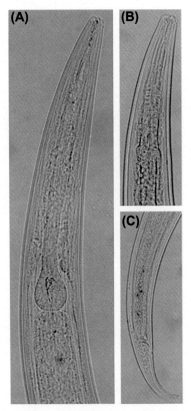

FIGURE 3.35 *Plectus* sp.: (A) pharyngeal region; (B) anterior end; (C) female posterior region.

and amphibians, but not predatory. The body is vermiform (except for most insect-parasitic and some semi/total-endoparasitic obese females). The length is generally not more than 3 mm. Males and juveniles of some genera lead a non-feeding, free-living life with reduced or complete absence of a stylet. The cuticle is two-layered: an epicuticle and an exocuticle, with or without ridges. Lateral fields have thickened cuticular markings, except in criconematid females with a thick cuticle and coarse annulations, sometimes with outgrowths in the form of spines and scales. Amphids have small fovea and pore- or slit-like apertures. Deirids are mostly present. Phasmids are small pore-like or large scutellum-like, and are absent in Hexatylina. The cephalic framework is lightly or heavily sclerotized, usually six-sectored, and six- or eight-sectored in Hexatylina. The oral aperture is pore- or slit-like. The stylet is composed of a conus, shaft, and basal knobs. Basal knobs are rarely absent. The pharynx is composed of a procorpus, a muscular median bulb with or without valve plates, a slender isthmus, and a glandular basal bulb or diverticulum. A pharyngo-intestinal valve is always present. A cardia may or may not be present. An intestinal lumen is not clear. The female genital system is monoprodelphic, didelphic-amphidelphic,

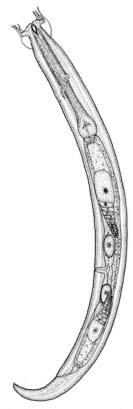

FIGURE 3.36 *Wilsonema* sp.: entire female.

or didelphic prodelphic; ovaries are generally outstretched. The male genital system is generally monorchic except for some sex-reversed males. Spicules are paired and cephalated. The gubernaculum is fixed or protrusible. The bursa is usually present, simple or lobed.

Type suborder

Tylenchina Chitwood in Chitwood and Chitwood, 1950*

Other suborders

Hexatylina Siddiqi, 1980
Criconematina Siddiqi, 1980*
Myenchina Siddiqi, 1980

Suborder Tylenchina Chitwood in Chitwood and Chitwood, 1950

Diagnosis: Tylenchida. The lip region is continuous or set off from the body contour. Cuticular striations are usually interrupted by lateral lines or ridges.

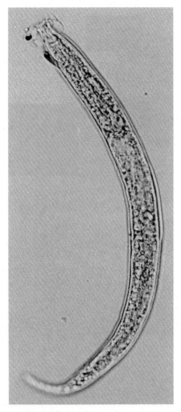

FIGURE 3.37   *Wilsonema* sp.: entire female.

The stoma is provided with a stylet. The pharynx in females is composed of a muscular precorpus or procorpus, a metacarpus/median bulb, and a glandular terminal/basal bulb or lobe. The dorsal gland nucleus opens in the procorpus. The median bulb usually has valve plates. The median and basal bulbs are connected by an isthmus. The nerve ring encircles the isthmus. The glandular terminal part of the pharynx may join the intestine or overlap the anterior part of intestine at different degrees. The female genital system is monodelphic or amphidelphic. Males usually have a bursa and paired spicules and gubernaculums. Phasmids or phasmid-like structures are present.

Type superfamily

Tylenchoidea Örley, 1880*

Other superfamilies

Dolichoroidea Chitwood in Chitwood and Chitwood, 1950*
Hoplolaimoidea Filipjev, 1934*
Anguinoidea Nicoll, 1935 (1926)*

Superfamily Tylenchoidea Örley, 1880

Diagnosis: Tylenchina. These are parasites of the subterranean and aerial parts of plants. The body is vermiform; sexual dimorphism in the anterior body region is absent. Deirids are almost always present. Phasmids are not detectable on the tail; a phasmid-like structure is present dorsal to the lateral field in the female near the vulva. The cuticle has fine or distinct transverse annulations. Longitudinal striations may be present. Cuticular annulations are not retrorse; they never have scales, spines, appendages, or a double cuticle. The cephalic region has distinct annulations, sometimes with longitudinal striae. The cephalic framework is hexaradiate, and lightly or heavily sclerotized. The oral opening is circular, oval, or slit-like. The stylet is generally small and the conus and shaft are usually of equal length. The shape and size of the basal knobs vary and are very rarely absent. The procorpus is muscular, cylindrical, or fusiform; the median bulb is usually valvulated and the isthmus is slender. The pharyngeal glands form the basal bulb or extend over the intestine as a lobe. The cardia is present and is greatly reduced in lobular forms.

The female genital system is monoprodelphic; the postvulval uterine sac is shorter than the body width or, rarely, is absent. The spermatheca is thin-walled, round, oval, or elongate, offset or axial. The testis is single and anteriorly outstretched, with the tip rarely reflexed. Spicules are paired and similar, cephalated, and ventrally arcuate; the gubernaculum is simple; and the bursa is adanal. The tail is similar between sexes: elongated and tapered, and usually filiform.

Type family

Tylenchidae Örley, 1880*

Other families

Atylenchidae Skarbilovich, 1959
Ecphyadophoridae Skarbilovich, 1959
Tylodoridae Paramonov, 1967

Family: Tylenchidae Örley, 1880

Diagnosis: The body is vermiform with a small to medium length (0.3–1.3 mm), rarely longer. The cuticle has transverse striations and may have longitudinal ridges. Lateral fields have two to four incisures. Deirids are present. Phasmids are dorso-sublateral postmedian. The lip region is rounded and elevated, provided with annulations, and rarely smooth. The labial framework is weakly sclerotized. The stylet is generally small (6–21 μm); the conus is usually less than half the length of the stylet, rarely with a distinct lumen; basal knobs are small, rounded, or absent. Amphidial apertures are variable: arc shaped, with a pore or slit-like. The oral disc is rounded and may or may not be elevated. The procorpus musculature is variable; the median bulb has valve plates that are moderately or strongly developed, spindle shaped, or oblong; the isthmus is long and slender; the basal bulb is glandular and pyriform, and rarely has a short overlapping lobe over the intestine. The cardia is small. The hemizonid and excretory pore are near each other. The nerve ring encircles the isthmus. The female genital system is monoprodelphic; the postvulval uterine sac is one body width long or less, and rarely is absent. Crustaformeria usually has a tricolumella. Spicules are small, slender, and arcuate; the bursa is adanal, simple or lobed. The tail is elongate conoid to long and filiform, and is similar in sexes. Members of this family are associates of algae, mosses, lichens, and plant roots.

Type genus

Tylenchus Bastian, 1865

Key characteristics of some commonly found genera of Tylenchidae

1. There are lateral fields, each with a single ridge 3
   There are lateral fields, each with double ridges 2
2. The stylet has a conus shorter than the shaft; the tail is not curved ventrally . . . .*Filenchus*
   The length of the conus and the shaft are almost equal; the tail is ventrally curved, sometimes hook-like . . . .*Tylenchus*
3. The lip region is elevated; the body is markedly narrow behind the vulval region . . . .*Malenchus*
   The lip region is not elevated; the body is uniformly constricted behind the vulval region; the vagina is straight; a postvulval uterine sac is present
   . . . .*Ottolenchus*
   The vagina is anteriorly directed; the postvulval uterine sac is absent . .4
4. The median bulb is strong and provided with valve plates . . .*Zanenchus*
   The median bulb is weak 5
5. Median valve plates are absent . . . . *Duosulcius*
6. Median valve plates are absent; the lateral field has three ridges
   . . . . *Boleodorus*
7. The median bulb is absent or very weak . . . . *Sakia*
8. The median bulb is strong and provided with valve plates; the stylet has basal knobs; the orifice of the dorsal pharyngeal gland is up to one stylet length behind the stylet base . . . .*Basiria*
9. The stylet is moderately developed; the cuticle has both longitudinal and transverse striations; a postvulval uterine sac is present . . . . *Coslenchus*
   The postvulval uterine sac is absent. 10
10. The male cloacal lips are tubular; the cuticle has only transverse striations
    . . . . *Aglenchus*

Genera recorded

Tylenchus Bastian, 1865
Filenchus Andrássy, 1954

Malenchus Andrássy, 1968
Basiria Siddiqi, 1959
Boleodorus Thorne, 1941

Genus *Tylenchus* Bastian, 1865 (Figs. 3.38 and 3.39)

Diagnosis: The body is small to medium sized, curved ventrad upon fixation. The cuticle has distinct annulations. The lateral fields each have four incisures. The lip region is continuous with the body; the cephalic framework has light or no sclerotization. Amphids have longitudinal slit-like apertures. The stylet length is 8–21 μm with a conus half or less but not less than one-third the stylet length; basal knobs are rounded, sloping posteriorly. The pharyngeal median bulb is oval and valvate, anterior to the middle of the pharynx; the basal bulb is pyriform. The excretory pore is usually opposite the basal bulb. The cardia is distinct. The vulva has a transverse slit. The postvulval uterine sac is about

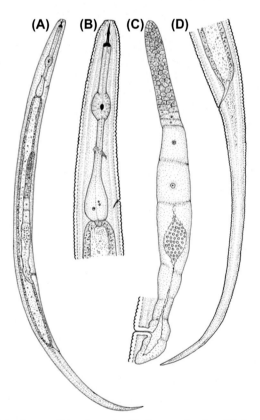

FIGURE 3.38   *Tylenchus* sp.: (A) entire male; (B) pharyngeal region; (C) female genital system; (D) female posterior region.

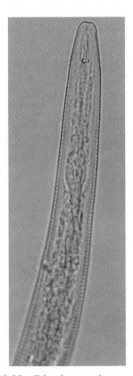

FIGURE 3.39   *Tylenchus* sp.: pharyngeal region.

a corresponding body width or less long. The spermatheca is oval to round and is offset. The ovary is outstretched. The tail is ventrally curved or often hooked, uniformly tapering to a pointed or rounded terminus. Spicules are arcuate, 13–25 μm long and cephalated. The gubernaculum is simple and fixed. The cloacal lips are slightly raised; the anterior lip is pointed, posterior rounded, and not tubular. The bursa is adanal, with crenate margins. The nematode is commonly found in most soils and feeds on algae, mosses, lichens, and plant roots.

Type species

Tylenchus davainei Bastian, 1865

Commonly found species of *Tylenchus* Bastian, 1865

T. davainei Bastian, 1865
T. arcuatus Siddiqi, 1963
T. elegans De Man, 1876

Genus *Filenchus* Andrássy, 1954 (Figs. 3.40 and 3.41)

Diagnosis: The body is small to medium sized (0.3–1.3 mm), straight to arcuate when relaxed. The cuticle has fine to moderate annulations. The lateral fields

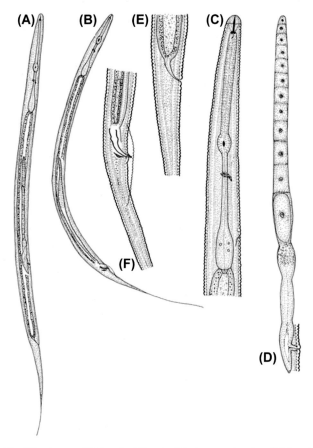

FIGURE 3.40   *Filenchus* sp.: (A) female; (B) male; (C) pharyngeal region; (D) female gonad; (E) female anal region; (F) male cloacal region.

each have four incisures. The amphidial apertures are straight longitudinal slits or clefts, labial in position. Deirids are present. The cephalic region is broadly rounded or rounded conoid, rarely truncated, continuous or slightly offset, with fine annulations. The labial disc is inconspicuous; the cephalic framework has light to moderate sclerotization. The stylet is weak or moderately developed; the conus is about one-third the stylet length, appearing solid anteriorly; knobs are rounded and distinct. The orifice of the dorsal pharyngeal gland is close behind the stylet base. The median bulb is muscular and oval to rounded, with valve plates; the basal bulb is pyriform; and the cardia is distinct. The vulva is postequatorial (55%–70%). The spermatheca is offset, lobular, and directed forward. The ovary is outstretched, usually with a single row of oocytes; a post-vulval uterine sac is present. The tail is usually filiform and straight; it may be elongate conoid, not ventrally curved or hooked. The bursa is adanal. Spicules are tylenchoid; the gubernaculum is cupped.

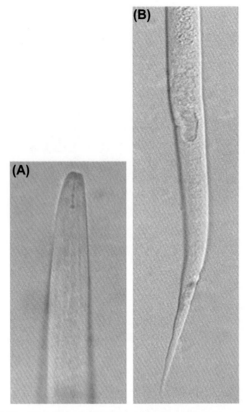

FIGURE 3.41 *Filenchus* sp.: (A) pharyngeal region; (B) male posterior region.

Type species

Filenchus vulgaris (Brzeski, 1963) Lownsbery and Lownsbery, 1983

Commonly Found Species of *Filenchus* Andrássy, 1954

F. vulgaris (Brzeski, 1963) Lownsbery and Lownsbery, 1983

Genus *Malenchus* Andrássy, 1968 (Figs. 3.42 and 3.43)

Diagnosis: The body is elongate fusiform, strongly tapering behind the vulva such that the width at the anus becomes about half that at the vulva. Annules are prominent. Lateral fields each have a single ridge marked by numerous fine longitudinal lines. The cephalic region is slightly flattened, with four or more annules. Amphids have ventrally curved, slit-like apertures. The procorpus is equal to or shorter than the isthmus; the median bulb is muscular, with refractive valve plates; the basal bulb is pyriform. The vulva is in a slight depression.

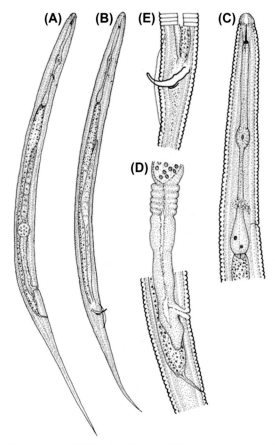

FIGURE 3.42 *Malenchus* sp.: (A) female; (B) male; (C) pharyngeal region; (D) female gonad; (E) male cloacal region.

The spermatheca is elongate, oval, or bilobed, and is offset. The vagina is straight, not sclerotized; a postvulval uterine sac is present. Phasmids are dorso-sublateral, about one body width anterior to the vulva. The tail is elongate conoid to a pointed or hooked tip. The bursa is adanal. Spicules are tylenchoid; the gubernaculum is fixed. The cloaca on a cone is formed by a depression in the body at the front and rear. The cloacal lips are narrow and pointed.

Type species

Malenchus machadoi (Andrássy, 1963) Andrássy, 1968

Commonly found species of *Malenchus* Andrássy, 1968

M. bryophilus Steiner, 1914
M. acarayensis Andrássy, 1968

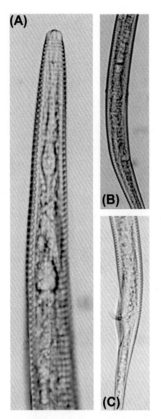

FIGURE 3.43 *Malenchus* sp.: (A) pharyngeal region; (B) female gonad; (C) male cloacal region.

Genus *Basiria* Siddiqi, 1959 (Figs. 3.44 and 3.45)

Diagnosis: The body is about 1 mm long or less, straight or curved ventrad upon relaxation. The cuticle has fine but distinct annules. The lateral fields each have four incisures; inner incisures are rarely indistinct. The amphidial apertures are arcuate, slit-like, or V shaped at the base of the lateral lips, posterior to the cephalic sensilla. The cephalic region is rounded and smooth; the framework is lightly sclerotized. The stylet is 9–13 μm long and slender, with small rounded basal knobs. The orifice of the dorsal pharyngeal gland is up to one stylet length behind the stylet base. The median bulb is generally poorly developed and muscular with refractive valve plates. The basal bulb pyriform is set off from the intestine. The cardia is distinct. The vulva is at 60%–70% of the body length from the anterior end; the spermatheca is lobed; the ovary is outstretched; and the postvulval uterine sac is shorter than one body width. The tail is elongate filiform with a clavate, rounded, pointed, or indented terminus. Phasmids are indistinct, dorso-sublateral, outside lateral fields. The bursa is adanal. Spicules are 14–24 μm long. The gubernaculum is simple and fixed.

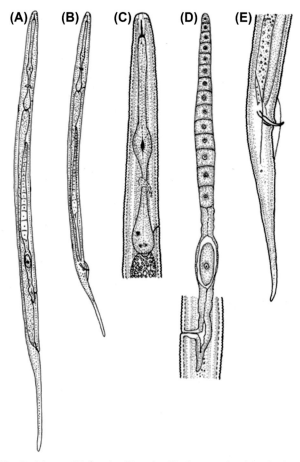

FIGURE 3.44   *Basiria* sp.: (A) female; (B) male; (C) pharyngeal region; (D) female gonad; (E) male posterior region.

Type species

Basiria graminophila Siddiqi, 1959

Commonly found species of *Basiria* Siddiqi, 1959

B. graminophila Siddiqi, 1959
B. duplexa Hagemeyer and Allen, 1952
B. gracilis Thorne, 1949

Genus *Boleodorus* Thorne, 1941 (Figs. 3.46 and 3.47)

Diagnosis: These nematodes are small with a body length usually less than 1 mm. The cuticle has fine annulations. The lateral fields each have four incisures.

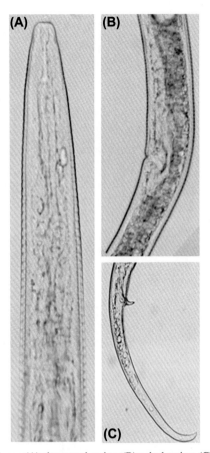

FIGURE 3.45 *Basiria* sp.: (A) pharyngeal region; (B) vulval region; (C) male posterior region.

The lip region is raised and conoid to rounded; the oral aperture can have a depression or not. The cephalic framework has weak or moderate sclerotization. Amphids have oval or crescent apertures. The stylet length is 8–10 µm; the conus is shorter than the shaft, usually one-third the total stylet length; the basal knobs are usually flanged and sometimes rounded. The median pharyngeal bulb is represented by fusiform swelling without musculature and valve plates. The length of the prometacorpus is usually greater than that of the postcorpus and isthmus combined; the basal bulb is pyriform. The vulva is postequatorial, usually at 59%–75% of the body from the anterior end. A short postvulval uterine sac is present. The tail is usually ventrally arcuate, rarely straight, and generally clavate. Spicules are slender; the gubernaculum is simple and trough shaped.

Type species

Boleodorus thylactus Thorne, 1941

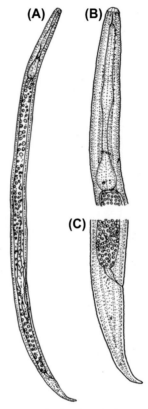

FIGURE 3.46   *Boleodorus* sp.: (A) entire female; (B) pharyngeal region; (C) posterior region.

Commonly found species of *Boleodorus* Thorne, 1941

B. thylactus Thorne, 1941
B. clavicaudatus Thorne, 1941
B. cynodonti Fotedar and Mahajan, 1974

Superfamily Dolichodoroidea Chitwood in Chitwood and Chitwood, 1950

Diagnosis: Marked sexual dimorphism in the anterior region is absent. The cuticle has prominent annulations. Each lateral field has two to six incisures, except Belonolaimus. Amphids are labial or just postlabial. Deirids are present or absent. Phasmids are on the tail or just preanal in females and extend into the bursa forming pseudoribs in males. The labial disc may or may not be prominent. The cephalic region is rounded, hexagonal, or four-lobed and

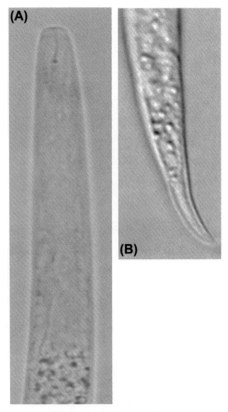

FIGURE 3.47 *Boleodorus* sp.: (A) pharyngeal region; (B) posterior region.

offset; annules are usually distinct; the framework has light or heavy sclerotization. The stylet length is variable, with knobs distinct except in Psilenchidae (absent). The dorsal pharyngeal gland opens near the basal knobs. The procorpus is cylindrical; the median bulb is oval or round and is provided with valve plates; the isthmus is slender; the postcorpus is small and bulb-like; and the dorsal pharyngeal gland may be enlarged, overlapping the anterior part of the intestine. The pharyngo-intestinal junction has a large or small valve (when the dorsal gland overlaps the intestine). The female genital system is amphidelphic except in Trophurinae (in which it is pseudomonoprodelphic). The vulva is median or submedian; the vulval opening is a transverse slit, sometimes oval or rarely round. The female tail is short to long and variously modified: conoid, cylindroid, subclavate, or elongate filiform. Spicules are cephalated, with or without distal flanges. The gubernaculum is simple, fixed, or modified, and protrusible. The male tail has the bursa enveloping the entire tail or adanal (Psilenchidae).

Type family

Dolichodoridae Chitwood in Chitwood and Chitwood, 1950*

Other family

Psilenchidae Paramonov, 1967*

Family Dolichodoridae Chitwood in Chitwood and Chitwood, 1950

Diagnosis: These nematodes are small to medium-sized (0.5–1.5 mm long). The cuticular annulations are strong (except in *Macrotrophurus*). Lateral fields each have one to six incisures. Deirids are present or absent. Phasmids are caudal and pore-like. The cephalic region is hexagonal or divided into four lobes. The labial disc is indistinct or distinct. Amphidial apertures are pore- or slit-like at lateral margins of the labial disc. The stylet is well developed, with basal knobs; the conus is not much shorter than the shaft. The pharyngeal glands are generally in the basal bulb abutting the intestine, or the dorsal gland may be enlarged, forming most of the lobe. Subventral glands are not enlarged. The female genital system is amphidelphic or rarely pseudomonoprodelphic. The vulva is median or submedian, transverse slit-like, and rarely transversely oval. The vulval lips are simple or modified. The vagina may be sclerotized. The spermathecae are well developed, usually with sperm. The spicules are strong and arcuate, with or without distal flanges. The gubernaculum is fixed or protrusible, without titilla. The bursa is simple or trilobed. The tail is dissimilar between sexes: long and filiform to spicate in females and short, conical, and pointed in males.

Type genus

Dolichodorus Cobb, 1914

Key characteristics of some commonly found genera of Dolichodoridae

1. There are lateral fields, each with three incisures; the cephalic region is prominently four-lobed; the female has a filiform or spicate tail . . . . .
   . . . . *Dolichodorus*
2. There are lateral fields, each with four incisures; the cephalic region is not prominently lobed; the female has a mammillate or obtuse tail . . . .
   . . . . *Neodolichodorus*
3. The lateral fields that are areolated, each with three incisures; the cephalic region has light sclerotization . . . . *Meiodorus*
4. The lateral fields are not areolated, each has four incisures; the cephalic region has heavy sclerotization; the vagina is sclerotized
   .. . . *Brachydorus*
5. The lateral fields each have four incisures; the cuticle has 12–20 longitudinal ridges; the bursa is notched at the tail tip . . . .
   . . . . *Dolichorhynchus*

6. The stylet is nearly 100 µm long; there are two gonads; the female tail terminus is abnormally thickened; the postrectal intestinal sac is absent . . . . *Macrotrophurus*

7. The stylet is less than 25 µm long; the gonad is single; the female tail terminus is broadly rounded; the postrectal intestinal sac is absent . . . . *Trophurus*

8. The cephalic annules are broken by six grooves; the tail terminus is of normal thickness. . . . .*Merlinius*

9. The cephalic annules are not broken by six grooves; the tail terminus is of abnormal thickness . . . . *Amplimerlinius*

10. The body cuticle has prominent longitudinal striations; deirids are absent; the vagina vera is squarish upon lateral view . . *Scutylenchus*

11. The lateral fields each have five incisures; the longitudinal ridges outside lateral fields are absent. . . . . *Quinisulcius*

12. The lateral fields each have four incisures, outer bands are areolated; the postanal intestinal sac is very large; the cuticle at the tail terminus is thickened. . . . . *Bitylenchus*

13. The cuticle has only transverse striations; the lateral fields each have four incisures; outer bands may be areolated; the gonads are amphidelphic . . . . *Tylenchorhynchus*

Genera recorded

Merlinius Siddiqi, 1970
Tylenchorhynchus Cobb, 1913
Dolichorhynchus Mulk and Jairajpuri, 1974
Quinisulcius Siddiqi, 1971

Genus *Merlinius* Siddiqi, 1970 (Figs. 3.48 and 3.49)

Diagnosis: The body is small, usually less than 1 mm, and straight to slightly curved ventrally in fixed materials. The lateral fields have six incisures each and usually are not areolated behind the pharyngeal region. The cephalic region is continuous or slight set off from the body by expansion. Annules in the lip region are interrupted by longitudinal striations or grooves. The stylet is moderately strong, not tubular near the tip. The median bulb is at the midpharyngeal length or slightly anterior. The vulva is slit-like and the epiptygma is indistinct. Female gonads are paired; spermatheca is two- to four-lobed. A postrectal intestinal sac is absent. The female tail is conoid to subcylindrical; the terminal cuticle is of normal thickness. The bursa is simple and moderately developed. Spicules are cylindroid, straight to slightly arcuate, with the distal end broadly rounded. The gubernaculum is fixed, simple, and trough-like.

Type species

Merlinius brevidens (Allen, 1955) Siddiqi, 1970

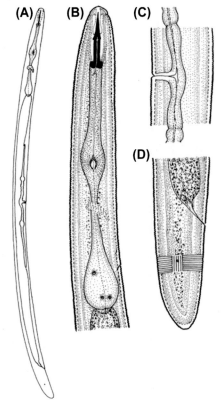

FIGURE 3.48   *Merlinius* sp.: (A) female; (B) pharyngeal region; (C) vulval region; (D) female posterior region.

Commonly found species of *Merlinius* Siddiqi, 1970

M. brevidens (Allen, 1955) Siddiqi, 1970
M. microdorus (Geraert, 1966) Siddiqi, 1970

Genus *Tylenchorhynchus* Cobb, 1913 (Figs. 3.50 and 3.51)

Diagnosis: The body is about 1 mm in length. The transverse annulations are distinct; longitudinal ridges are absent. Lateral fields have four incisures; outer bands may be areolated. The lip region is continuous or offset. The labial disc is indistinct. The cephalic framework has light to moderate sclerotization. The stylet is well developed; the anterior conus appears solid; knobs are prominent. The median bulb is oval to round; the basal bulb is offset from the intestine. The vulva is median, the spermatheca is axial, and ovaries are paired. The female tail is variable: conoid with a bluntly rounded terminus to cylindrical or clavate with a rounded terminus. Male tails are enveloped completely with bursa. Spicules are flanged with a narrow terminus that is indented or pointed. The gubernaculum is rodlike, protrusible, and about half the spicules' length.

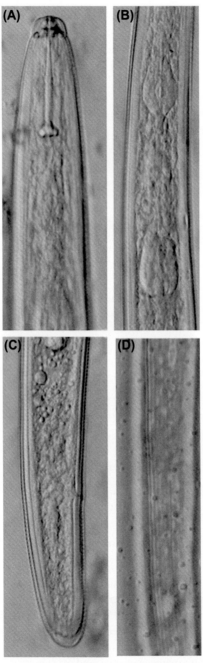

FIGURE 3.49 *Merlinius* sp.: (A) stomal region; (B) pharyngeal region; (C) female posterior region; (D) lateral field.

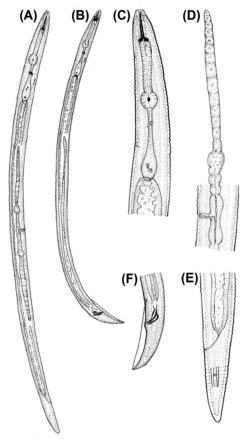

FIGURE 3.50 *Tylenchorhynchus* sp.: (A) female; (B) male; (C) pharyngeal region; (D) female genital system (anterior); (E) female posterior region; (F) male posterior region.

Type species

Tylenchorhynchus cylindricus Cobb, 1913

Commonly found species of *Tylenchorhynchus* Cobb, 1913

T. brevilineatus Williams, 1960
T. striatus Allen, 1955
T. nudus Allen, 1955
T. martini Fielding, 1956

Genus *Dolichorhynchus* Mulk and Jairajpuri, 1974
(Figs. 3.52 and 3.53)

Diagnosis: The small nematodes are usually less than 1 mm. They are arcuate ventrally or more strongly curved upon fixation. The cuticle has deep transverse

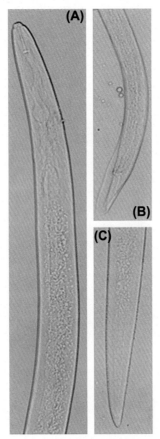

FIGURE 3.51 *Tylenchorhynchus* sp.: (A) pharyngeal region; (B) male posterior region; (C) female posterior region.

striations and 12–20 prominent longitudinal lamellae or ridges. The lateral fields have four lines that are variously areolated along the entire length. The lip region is continuous or offset. The labial disc is indistinct. The cephalic framework has light to moderate sclerotization. The stylet is well developed; the conus appears solid in the anterior third. The median bulb of the pharynx is oval or rounded, with well-developed valve plates. The basal bulb is set off from the intestine. The vulva is equatorial or postequatorial; lateral membranes are present or absent. The female tail is conoid to subcylindroid, ending in a rounded, lobe-like, hyaline terminus. The male tail has a large bursa that may or may not be notched. Spicules are provided with distal flanges. The gubernaculum is large and protrusible.

Type species

Dolichorhynchus phaseoli (Sethi and Swarup, 1968) Mulk and Jairajpuri, 1974

Commonly found species of *Dolichorhynchus* Mulk and Jairajpuri, 1974

D. phaseoli (Sethi and Swarup, 1968) Mulk and Jairajpuri, 1974

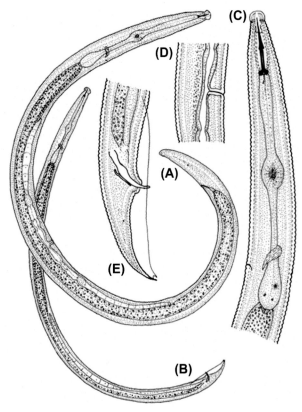

FIGURE 3.52 *Dolichorhynchus* sp.: (A) female; (B) male; (C) pharyngeal region; (D) vulval region; (E) male posterior region.

Genus *Quinisulcius* Siddiqi, 1971 (Figs. 3.54 and 3.55)

Diagnosis: The body is small, usually less than 1 mm, and strongly curved ventrally in fixed materials. The cuticle has prominent annulations. Longitudinal ridges or lamellae are absent. Each lateral field has five incisures, which may be smooth or areolated. The lip region is rounded, set off from the body, with fine annulations. The labial disc is indistinct and the framework is lightly or moderately sclerotized. The stylet is 12–24 μm long; the conus appears solid anteriorly; the basal knobs are rounded, sloping backward or directed anteriorly. The pharyngeal median bulb is well developed, oval, not set off from the precorpus. The basal bulb is set off from the intestine. The cardia is well developed. The vulva is just postequatorial at 51%–59%; vulval lips are not modified. Gonads are paired and the spermatheca is round, axial, or slightly offset. Females have a conoid, ventrally arcuate tail. The terminal annule of the tail is enlarged, smooth, or striated. Males are rare. The bursa is simple, enveloping the entire tail. Spicules are arcuate; distal flanges are poorly developed. The gubernaculum has a dorsally directed proximal end.

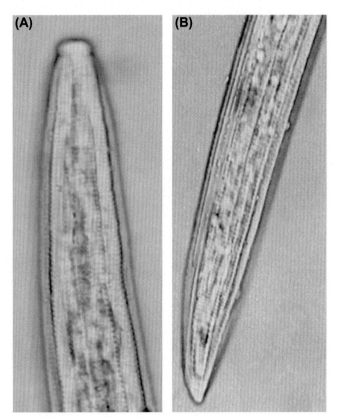

FIGURE 3.53 *Dolichorhynchus* sp.: (A) pharyngeal region; (B) female posterior region.

Type species

Quinisulcius capitatus (Allen, 1955) Siddiqi, 1971

Commonly found species of *Quinisulcius* Siddiqi, 1971

Q. capitatus (Allen, 1955) Siddiqi, 1971
Q. acutus (Allen, 1955) Siddiqi, 1971
Q. curvus (Williams, 1960) Siddiqi, 1971

Family Psilenchidae Paramonov, 1967

Diagnosis: The nematodes are small to medium-sized. The cuticle has prominent annulations. The lateral fields have four incisures each. Amphids have indistinct and pore-like or distinct and slit-like apertures. Phasmids are distinct and pore-like on the tail. The stylet is slender; the conus is shorter than the shaft; basal knobs are either present or absent. The median bulb of the pharynx

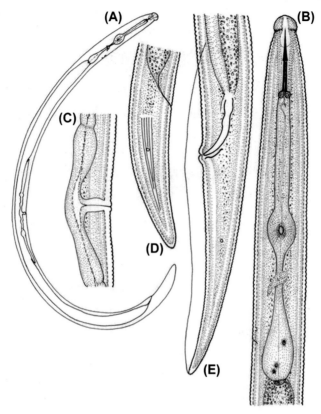

FIGURE 3.54 *Quinisulcius* sp.: (A) female; (B) pharyngeal region; (C) vulval region; (D) female posterior region; (E) male posterior region.

is muscular and valvate; the basal bulb is set off from the intestine; the cardia is prominent. The vulva is equatorial or subequatorial; the lateral membrane may be present or absent. The female gonads are paired and outstretched. The tail is filiform or elongate conoid, and is similar in both sexes. Spicules are tylenchoid and cephalated. The gubernaculum is simple and fixed. The bursa is simple and adanal.

Type genus

Psilenchus De Man, 1921

Key characteristics of some commonly found genera of Psilenchidae

1. Stylet has no knobs; cephalic region is smooth; amphidial apertures are postlabial . . . . *Psilenchus*
2. Stylet has no knobs; cephalic region is annulated; amphidial apertures are labial. . . . . *Atetylenchus*

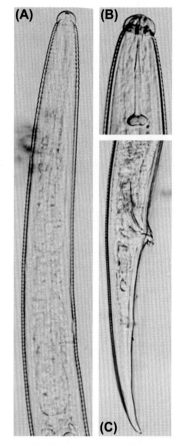

FIGURE 3.55   *Quinisulcius* sp.: (A) pharyngeal region; (B) anterior end; (C) male posterior region.

Genera recorded

Psilenchus De Man, 1921
Atetylenchus Khan, 1973

Genus *Psilenchus* De Man, 1921 (Figs. 3.56 and 3.57)

Diagnosis: These are medium-sized nematodes (L=0.7–1.7 mm). The body is ventrally curved upon fixation. Lateral fields each have four incisures; the inner two may be indistinct or absent. Amphids have slit-like apertures at the base of the lateral lip areas. Phasmids are distinct in the anterior half of the tail. The cephalic region is elevated, rounded or conoid, and smooth or striated. The cephalic framework is weakly sclerotized. The stylet is 10–24 μm long and cylindrical; the conus is shorter than the shaft; and basal knobs are absent. The median bulb is muscular and valvate, usually behind the middle of the pharynx.

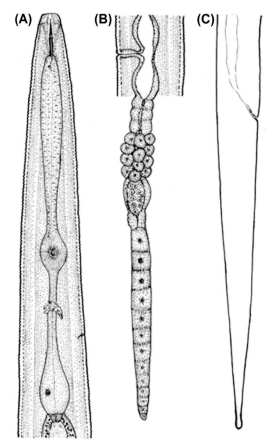

FIGURE 3.56 *Psilenchus* sp.: (A) pharyngeal region; (B) reproductive system (anterior); (C) tail region.

The basal bulb is small, glandular, and pyriform. The vulva is near the midbody; epiptygma or lateral membranes are absent. Female gonads are paired; ovaries are outstretched. The tail is elongated with a clavate or nonclavate rounded tip. The bursa is adanal. Spicules are tylenchoid, cephalated, and 25–33 μm long. The gubernaculum is simple and trough shaped. The nematode is commonly found in most soils. It feeds on algae, mosses, lichens, and plant roots.

Type species

Psilenchus hilarulus De Man, 1921

Commonly found species of *Psilenchus* De Man, 1921

P. hilarulus De Man, 1921
P. hilarus Siddiqi, 1963

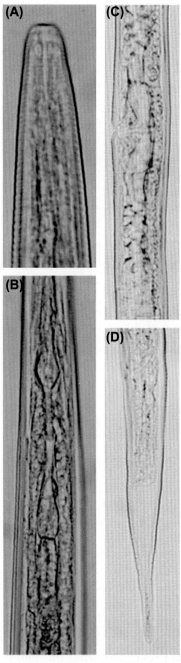

FIGURE 3.57 *Psilenchus* sp.: (A) stomal region; (B) pharyngeal region; (C) vulval region; (D) posterior region.

Genus *Atetylenchus* Khan, 1973 (Figs. 3.58 and 3.59)

Diagnosis: These are medium-sized nematodes with a body length usually less than 1.5 mm. The body is ventrally curved upon fixation. The lateral fields each have four incisures; areolations are absent. Amphids have indistinct apertures near the oral opening. Phasmids are minute and pore-like. The cephalic region is continuous with a body contour that is rounded and striated. The cephalic framework is weakly sclerotized. The stylet is cylindrical; basal knobs are absent. The median bulb is muscular and valvate, usually anterior to the middle of the pharynx. The basal bulb is small, glandular, and pyriform. The vulva is near the midbody. The female gonads are paired; ovaries are outstretched. The tail is elongated and ventrally curved, tapering to a finely rounded terminus. The terminal hyaline portion of the tail is inconspicuous. The bursa is adanal. Spicules are tylenchoid and cephalated. The gubernaculum is fixed, simple, and trough shaped.

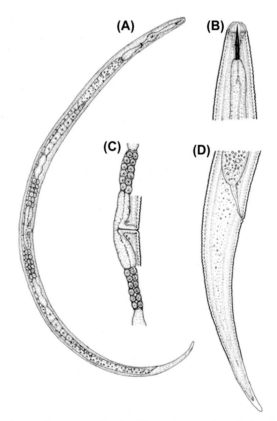

FIGURE 3.58 *Atetylenchus* sp.: (A) female; (B) anterior end; (C) vulval region; (D) posterior region.

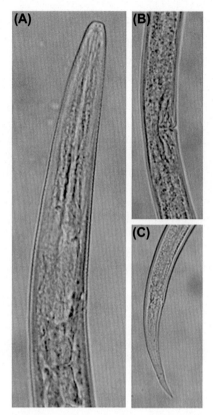

FIGURE 3.59   *Atetylenchus* sp.: (A) anterior end; (B) vulval region; (C) posterior region.

Type species

Atetylenchus abulbosus (Thorne, 1949) Khan, 1973

Commonly found species of *Atetylenchus* Khan, 1973

A. abulbosus (Thorne, 1949) Khan, 1973
A. graminis (Bajaj et al., 1982) Siddiqi 1986

Superfamily Hoplolaimoidea Filipjev, 1934

Diagnosis: The body size ranges from 0.5 to 2.0 mm. Sexual dimorphism is present in the anterior cephalic region. The cuticle has strong annulations; longitudinal ridges are absent. Lateral fields typically have four lines but they may be reduced toward the extremities. The lip region is elevated, strongly sclerotized, and less developed in males. The stylet is strongly developed; the conus and shaft are of almost equal length. The basal knobs are well developed, round or indented, and anchor or tulip shaped. The dorsal pharyngeal gland orifice is slightly or considerably behind the basal knobs. The median bulb is strongly developed. The pharyngeal glands generally overlap the anterior part of the intestine. The subventral glands are enlarged. The pharyngo-intestinal junction is small and triangular. The female genital system is amphidelphic with outstretched ovaries or, rarely, monoprodelphic with a degenerated posterior gonad or a postvulval uterine sac. Epiptygma or vulval flaps are conspicuous or sometimes inconspicuous. The female tail is elongated and conoid, short and conoid, or absent. The anterior end in males may be less developed than in females. The spicules are straight or arcuate and cephalated; they may or may not be flanged. The gubernaculum is fixed or protrusible, with or without titilla and telamon. The tail is short, generally with a hyaline terminus. The bursa is terminal, subterminal, or absent; the phasmidial pseudoribs are absent.

Type family

Hoplolaimidae Filipjev, 1934*

Other families

Heteroderidae Filipjev and Schuurmans Stekhoven, 1941
Meloidogynidae Skarbilovich, 1959
Nacobbidae Chitwood in Chitwood and Chitwood, 1950
Pratylenchidae Thorne, 1949*
Rotylenchulidae Husain and Khan, 1957

Family Hoplolaimidae Filipjev, 1934

Diagnosis: These nematodes are small to medium-sized (0.5–1.5 mm). The female is vermiform to kidney shaped. Sexual dimorphism is present in the cephalic region. Lateral fields typically have four incisures and are rarely regressed. Deirids are absent. The phasmids are either small, with pore-like apertures near or a little anterior to the anus, or large and scutellum-like, near the anus or much anterior to it anywhere on the body behind the pharyngeal region; they are rarely absent. The lip region is elevated and high arched; the labial disc is usually distinct. The cephalic framework is strongly sclerotized. The stylet is strong with rounded, indented, and sometimes anchor- or tulip-shaped knobs. The median bulb is strong and rounded. The pharyngeal glands usually overlap the anterior intestine. The pharyngo-intestinal junction is a small triangular structure. The female gonads are paired; they are rarely single. The genital branches are outstretched or rarely reflexed. The posterior branch may be degenerated or reduced to a postvulval uterine sac. The female tail is short and rounded or conical. Secondary sexual dimorphism is present. Males have less developed anterior ends than do females; sometimes they are degenerated and nonfunctional. The bursa is large and envelops the tail; it is rarely short and subterminal. The spicules are slender to robust and straight to arcuate, with distal flanges. The gubernaculum is fixed or protrusible. The capitulum is either present or absent.

Type genus

Hoplolaimus Daday, 1905

Key characteristics of some commonly found genera of Hoplolaimidae

1. The body is almost straight or slightly ventrally curved; phasmids are large and scutellum-like, not opposite each other; the stylet has tulip-shaped knobs .... *Hoplolaimus*
2. Phasmids are large and scutellum-like, located near the anal region, nearly or exactly opposite each other; the stylet has rounded knobs. . ... *Scutellonema*
3. The cephalic region is offset, annulated; the pharyngeal glands overlap the intestine dorsally or subdorsally; the phasmids are pore-like . . . . *Rotylenchus*
4. The cephalic region is continuous; the pharyngeal glands overlap the intestine ventrally or ventrolaterally; the phasmids are pore-like 5
5. Both gonads are functional. .... *Helicotylenchus*
   The posterior ovary is nonfunctional or absent .... *Rotylenchoides*

Genera recorded

Scutellonema Andrássy, 1958
Rotylenchus Filipjev, 1936
Helicotylenchus Steiner, 1945
Rotylenchoides Whitehead, 1958

Genus *Scutellonema* Andrássy, 1958 (Figs. 3.60 and 3.61)

Diagnosis: The body is small to medium-sized (0.5–1.3 mm). The lip region is continuous with the body or is set off; it is rounded, flattened, or truncated, with or without annulations. The basal lip annule does or does not have longitudinal striations. The lateral fields have four incisures each, with areolations anteriorly, generally at the level of the phasmids, sometimes throughout the entire length. The stylet is well developed with rounded, anteriorly flattened basal knobs. The pharyngeal glands overlap the intestine dorsally and laterally. The female genital system is amphidelphic; both ovaries are equally developed. The tail is short, usually less than one anal body in diameter. The phasmids are large (scutella) and opposite or almost opposite, at the level of the anus. The bursa completely envelops the tail; the margins are crenate and sometimes lobed.

Type species

Scutellonema bradys (Steiner and Le Hew, 1933) Andrássy, 1958

Commonly found species of *Scutellonema* Andrássy, 1958

S. bradys (Steiner and Le Hew, 1933) Andrássy, 1958
S. orientale (Rashid and Khan, 1974) Mattaar and Loof, 1984
S. brachyurum (Steiner, 1938) Andrássy, 1958
S. clathricaudatum Whitehead, 1959

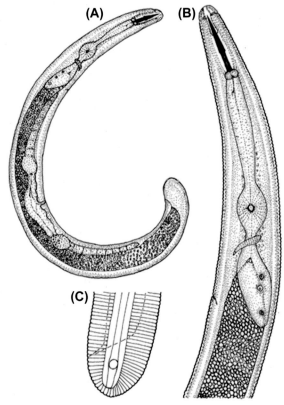

FIGURE 3.60 *Scutellonema* sp.: (A) female; (B) pharyngeal region; (C) posterior region showing scutellum.

Genus *Rotylenchus* Filipjev, 1936 (Figs 3.62 and 3.63)

Diagnosis: The body is a spiral to a closed C shape upon relaxation. The lip region is rounded and continuous with the body contour or set off from it, with or without longitudinal striae on the basal lip annules. The lateral fields have four lines each. The phasmids are pore-like and are present at the same level, usually near the anal region. The cephalic framework is moderately developed. The stylet is moderately developed; the knobs are rounded. The dorsal pharyngeal gland opens close to the stylet base or slightly posterior to it. The pharyngeal glands overlap the intestine dorsally, subdorsally, or laterally. The female genital system is amphidelphic; both ovaries are equally developed. The spicules are robust and flanged distally. The gubernaculum is not fixed. The bursa envelops the tail and is not indented terminally.

Type species

Rotylenchus robustus Filipjev, 1936.

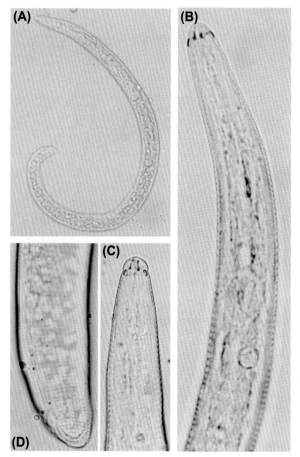

FIGURE 3.61   *Scutellonema* sp.: (A) female; (B) pharyngeal region; (C) anterior region; (D) posterior region.

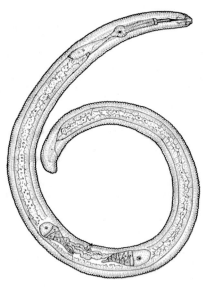

FIGURE 3.62   *Rotylenchus* sp.: female.

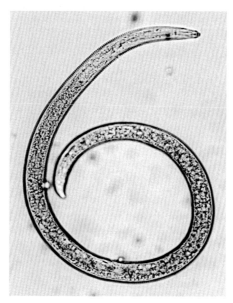

FIGURE 3.63 *Rotylenchus* sp.: female.

Commonly found species of *Rotylenchus* Filipjev, 1936

R. robustus Filipjev, 1936
R. buxophilus Golden, 1956
R. elegans (Khan and Khan, 1982) Fortuner, 1987

Genus *Helicotylenchus* Steiner, 1945 (Figs. 3.64 and 3.65)

Diagnosis: The nematodes are small to medium-sized (0.4–1.2 mm). The body is strongly spirally coiled and very rarely arcuate. The lateral fields have four incisures each. The cephalic region is low or raised, continuous or slightly set off, and rounded or anteriorly flattened; it does or does not have annulations but it never has longitudinal indentations or striations. The cephalic framework is moderate. The stylet is robust, three to four cephalic region widths long. The dorsal pharyngeal gland orifice is at one-quarter to about half the length of the stylet behind the stylet base. The pharyngeal glands extend slightly over the intestine; the longest extension is on the ventral side. The pharyngo-intestinal junction has a small cuticular valve. Both gonads are well developed. The female tail is short, hemispherical, and dorsally convex conoid, with or without a ventral or terminal projection. The phasmids are small and near the anus. The gubernaculum is fixed and trough shaped or rodlike. The male tail is short and conical, with a distinct terminal hyaline portion. The bursa envelops the entire tail tip and is rarely subterminal.

FIGURE 3.64 *Helicotylenchus* sp.: entire female.

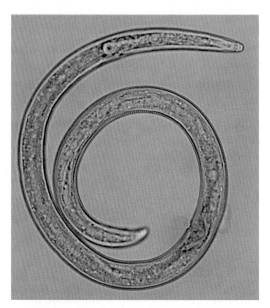

FIGURE 3.65 *Helicotylenchus* sp.: entire female.

Type species

Helicotylenchus dihystera (Cobb, 1893) Sher, 1961

Commonly found species of *Helicotylenchus* Steiner, 1945

H. dihystera (Cobb, 1893) Sher, 1961
H. multicinctus (Cobb, 1893) Golden, 1958
H. pseudorobustus (Steiner, 1914) Golden, 1956
H. digonicus Perry, 1959

Genus *Rotylenchoides* Whitehead, 1958 (Figs. 3.66 and 3.67)

Diagnosis: These are small nematodes (0.37–0.58 mm). The body is arcuate or more curved. The lateral fields have four incisures each; areolations are absent behind the pharynx. The cephalic region is low, continuous, and rounded with annulations. The cephalic framework has heavy sclerotization. The stylet is robust, about three cephalic region widths long. The dorsal pharyngeal gland orifice is at less than half the length of the stylet behind the stylet base. The pharyngeal glands extend slightly over the intestine. The pharyngo-intestinal junction has a small cuticular valve. The posterior branch of the female genital system is nonfunctional or reduced, or is represented by a short postvulval uterine sac. The vulva is located posteriorly. The tail is short, hemispherical, and dorsally convex conoid. The phasmids

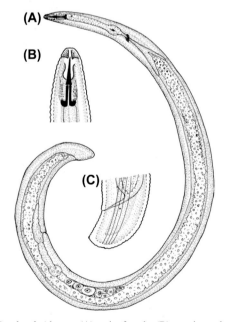

FIGURE 3.66 *Rotylenchoides* sp.: (A) entire female; (B) anterior end; (C) posterior end.

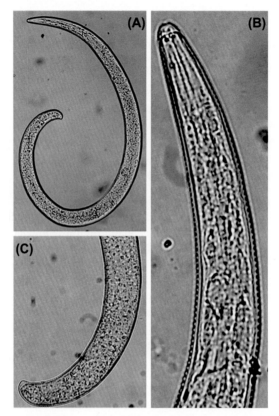

FIGURE 3.67 *Rotylenchoides* sp. (A) Entire female; (B) Pharyngeal region; (C) Posterior end.

are small and near the anus. Males have distally slender spicules and a fixed gubernaculum.

Type species

Rotylenchoides brevis Whitehead, 1958

Commonly found species of *Rotylenchoides* Whitehead, 1958

R. brevis Whitehead, 1958
R. affinis Luc, 1960
R. cheni Zhu, Lan, Hu, Yang and Wang, 1991
R. variocaudatus Luc, 1960

Family Pratylenchidae Thorne, 1949

Diagnosis: These are vermiform nematodes. The mature females are rarely swollen. Sexual dimorphism is frequent. The cuticle has prominent annulations.

The lateral fields have four to six incisures each, sometimes with areolations behind the pharynx. Deirids are absent (except in *Pratylenchoides*). The phasmids are pore-like. The lip region is low; the labial framework is heavily sclerotized. The stylet is strong, usually less than three lip region widths in length (except in *Hirschmanniella*); the conus is almost as long as the shaft; the basal knobs are fairly developed and rounded or anteriorly flat. The median bulb is strongly developed and round or oval, and is provided with strong valve plates. The pharyngeal glands extend over the intestine (with the exception of a few species of *Pratylenchoides*). The valve at the pharyngo-intestinal junction is generally not well developed. Females have an amphidelphic or monoprodelphic genital system. The vulval flap or epiptygma is absent. The spicules are arcuate and cephalated, opening subterminal on the ventral or dorsal side. The gubernaculum is fixed and simple or protrusible with titilla and telamon. The hypoptygma is present or absent. The bursa is terminal or subterminal. The tail is twice as long as the anal body diameter (except in *Nacobbus*).

With the exception of *Nacobbus*, in which the females are sedentary, forming galls on roots, representatives of this family are generally obligate migratory endoparasites.

Type genus

Pratylenchus Filipjev, 1936

Key characteristics of some commonly found genera of Pratylenchidae

1. No marked sexual dimorphism in the anterior region; only the anterior gonad is functional; the bursa is terminal . . . . *Pratylenchus*
2. Both gonads are functional; the bursa is subterminal; the tail terminus usually has a mucro . . . . *Hirschmanniella*
3. Marked sexual dimorphism in the anterior region; all pharyngeal gland nuclei are posterior to the pharyngo-intestinal junction; both gonads are functional . . . . *Radopholus*
4. Marked sexual dimorphism in the anterior region; minimum of one pharyngeal gland nucleus anterior to the pharyngo-intestinal junction; both gonads are functional . . . . *Pratylenchoides*

Genus recorded

Pratylenchus Filipjev, 1936

Genus *Pratylenchus* Filipjev, 1936 (Figs. 3.68 and 3.69)

Diagnosis: The body is short, less than 1 mm. Sexual dimorphism is absent in the anterior region. The lateral fields have four to six incisures each. Median oblique markings are occasionally present in the lateral fields. Deirids are absent. The lip region is continuous with the body contour; the labial disc is inconspicuous. The cephalic region is low and anteriorly flattened or rarely

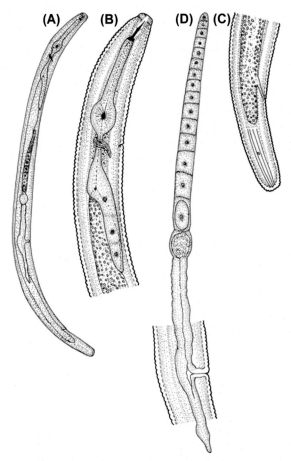

FIGURE 3.68 *Pratylenchus* sp.: (A) female; (B) pharyngeal region; (C) posterior region; (D) female gonad.

rounded; the cephalic framework is strongly sclerotized. The amphidial apertures are pore-like, near the labial disc or indistinct. The stylet is well developed, about 20 μm long; the basal knobs are round, anteriorly flat, or indented. The pharyngeal median bulb is strongly muscular and oval to round. The pharyngeal glands are generally less than two body widths long and overlap the intestine, mostly ventrally. The vulva is postequatorial at 70%–80% of the body from the anterior end. The female genital system is pseudomonoprodelphic. A postvulval uterine sac is present; rudiments of the posterior ovary may be present or absent. The spermatheca is large, round, and axial. The female tail is subcylindrical or conoid, two to three ABDs long. The tail terminus is smooth or annulated. The spicules are arcuate, with a subterminal pore on the dorsal side. The gubernaculum is fixed, simple, and trough shaped. The bursa encloses the tail terminus.

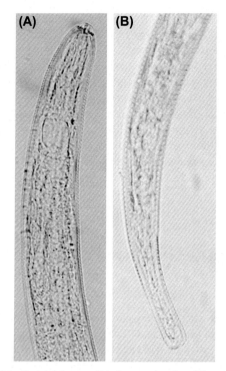

FIGURE 3.69 *Pratylenchus* sp.: (A) pharyngeal region; (B) posterior region.

Type species

Pratylenchus pratensis (De Man, 1880) Filipjev, 1936

Commonly found species of *Pratylenchus* Filipjev, 1936

P. pratensis (De Man, 1880) Filipjev, 1936
P. coffeae (Zimmerman, 1898) Filipjev and Stekhoven, 1941
P. brachyurus (Godfrey, 1929) Filipjev and Stekhoven, 1941
P. penetrans (Cobb, 1917) Filipjev and Stekhoven, 1941
P. scribneri Steiner, 1943
P. thornei Sher and Allen, 1953
P. zeae Graham, 1951

Superfamily Anguinoidea Nicoll, 1935 (1926)

Diagnosis: These are small to large-sized nematodes. Adults of some exceptional species may be obese. Marked sexual dimorphism is absent in the anterior region. The cuticle is finely, transversely striated. The lateral fields are smooth or have four, six or more incisures. The lip region is low and smooth. The amphidial apertures are generally indistinct. The stylet is small, less than

15 μm in length; the basal knobs are small and rounded. The dorsal esophageal gland opens near the stylet base. The median bulb of the pharynx is either present or absent. The esophageal glands are generally in the basal bulb; the dorsal gland sometimes extends over the intestine. The cardia is absent. The female reproductive system is monoprodelphic. The vulva is located posteriorly. The postvulval uterine sac is usually more than one vulval body width (except it is absent in *Diptenchus*). The spicules are large and the gubernaculum is simple and trough-shaped; it is rarely absent. The size of the bursa varies from adanal to subterminal; it is rarely terminal. The tail is usually similar in sexes, or dissimilar in those with a terminal bursa.

Type family

Anguinidae Nicoll, 1935 (1926)*

Other family

Sychnotylenchidae Paramonov, 1967

Family Anguinidae Nicoll, 1935 (1926)

Diagnosis: These are small to large nematodes and either slender or obese. The lip region is low and smooth. The amphidial apertures are generally indistinct. A valvulated median bulb of the pharynx is either present or absent. The esophageal glands are generally in the basal bulb; the dorsal gland sometimes extends over the intestine. The cardia is absent. The female reproductive system is monoprodelphic. The ovary is outstretched or has one or two flexures. The vulva is located posteriorly. The postvulval uterine sac is usually more than one vulval body width (except it is absent in *Diptenchus*). The spicules are large. The gubernaculum is present.

Type genus

Anguina Scopoli, 1777

Key characteristics of some commonly found genera of Anguinidae

1. The female is obese and spirally curved; the median pharyngeal bulb is strongly developed; cells of the columella are arranged in more than four rows . . . . *Anguina*
2. The dorsal gland of the esophagus is located anterior to the esophagus–intestinal junction; the postvulval uterine sac is absent. . . *Diptenchus*
3. The median bulb of the esophagus is muscular and distinct. . *Ditylenchus*
4. The median bulb of the esophagus is not muscular and indistinct . . . . . .
. . . . *Nothotylenchus*

Genus recorded

Ditylenchus Filipjev, 1936

Ditylenchus Filipjev, 1936 (Figs. 3.70 and 3.71)

Diagnosis: These are small to medium-sized nematodes, usually less than 1.5 mm in length. The body is slightly ventrally curved upon fixation. The lateral fields each have four to six incisures or sometimes more. The median esophageal bulb may or may not be muscular, with or without a valve plate. The esophageal gland nuclei is always anterior to the esophago-intestinal junction and does not form a long lobe over the intestine. The ovary extends anteriorly. The vagina is more or less perpendicular to the body axis and is not directed anteriorly. A postvulval uterine sac is present. Males have arcuate spicules, a trough-shaped gubernaculum, and an adanal to subterminal bursa. The bursa never reaches the tail tip. The tail in both sexes is elongate conoid to subcylindrical or occasionally filiform.

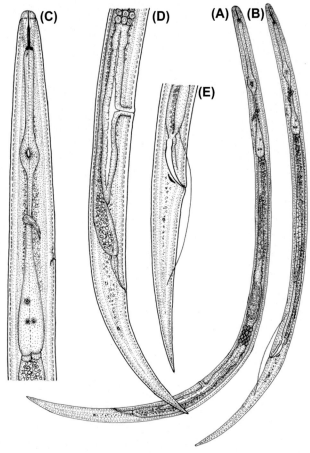

FIGURE 3.70 *Ditylenchus* sp.: (A) female; (B) male; (C) pharynx; (D) female posterior region; (E) male posterior region.

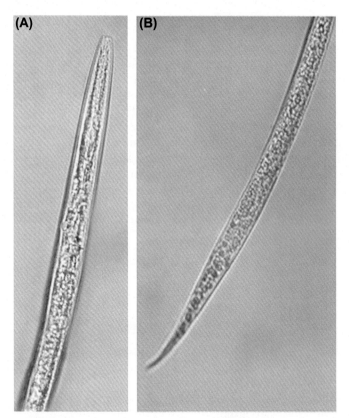

FIGURE 3.71 *Ditylenchus* sp.: (A) pharynx; (B) female posterior region.

Type species

Ditylenchus dipsaci (Kühn, 1857) Filipjev, 1936

Commonly found species of *Ditylenchus* Filipjev, 1936

D. angustus (Butler, 1913) Filipjev, 1936
D. angustus (Micoletzky, 1925) Filipjev, 1936
D. destructor Thorne, 1945
D. dipsaci (Kühn, 1857) Filipjev, 1936

Suborder Criconematina Siddiqi, 1980

Diagnosis: The males and juveniles are vermiform; the females are vermiform or obese (Tylenchuloidea) and usually less than 1 mm. Sexual dimorphism in the anterior region is pronounced; males have a degenerated pharynx. The stylet is degenerated or even absent. The cuticle is thin with fine annulations or thick with variously modified annulations (retrorse annules, spines, and scales). An extracuticular body sheath is either present or absent. Lateral fields are largely absent. Phasmids are always absent. The lip region is smooth or annulated. The cephalic framework is lightly or heavily sclerotized. The stylet is usually long; the length of the conus varies; the shaft is generally 8–10 µm long. The basal knobs are large and typically anchor shaped in larger ones. The precorpus and metacorpus are well developed and fused together; the isthmus is very short or amalgamated with the basal bulb. The basal bulb is usually small. The cardia is small or indistinct. The excretory pore is in the pharyngeal region or further posterior. The female genital system is monoprodelphic; the postvulval uterine sac is absent. The testis is single and usually obliterated in adults. The spicules are usually long and variously shaped. The gubernaculum is simple, fixed, rodlike, or slight curved. The caudal alae is absent or weakly developed; sometimes it envelops the tail tip.

Type superfamily

Criconematoidea Taylor, 1936*

Other superfamilies

Hemicycliophoroidea Skarbilovich, 1959
Tylenchuloidea Skarbilovich, 1947*

Superfamily Criconematoidea Taylor, 1936

Diagnosis: The nematodes are usually small. The females have a sausage- or spindle-shaped or cylindrical body. Sexual dimorphism is pronounced. The cuticle in males is thinner than that in females and always has rounded annulations. Incisures are present in the lateral fields. The stylet is absent. The pharynx lacks a distinct structure. The females have a thick cuticle provided with retrorse or rounded coarse annulations. Scales or spines or other associated structures are either present or absent. The conus is much longer than the shaft; the basal knobs are usually anchor shaped. The precorpus is broad, expanding posteriorly to fuse with the metacorpus; the pharyngeal lumen is spiral; the valve plates are elongated. The isthmus is short and broad. The basal bulb is small and set off from the intestine. The intestine has no clear lumen. The vulva is posterior; the ovary is outstretched. The vagina is directed anteriorly and lacks a postvulval uterine sac. The spicules are elongate setose and cephalated, almost straight to arcuate. The gubernaculum is simple and fixed. The bursa is absent, adanal, subterminal, or sometimes terminal.

Type and only family

Criconematidae Taylor, 1936*

Family Criconematidae Taylor, 1936

Diagnosis: The body is short, usually less than 1 mm, with marked sexual dimorphism.

For females, the body is sausage shaped or cylindrical. The cuticle is thick, with 33–200 annules prominent and retrorse, with or without scales or spines; lateral fields are absent or marked by irregular body annules and/or superficial longitudinal lines. Body annules are retrorse; they may or may not be provided with lobation, crenation, scales, or spines, or they may be rounded and may or may not be covered with an extra layer of cuticle. The lip region is variously shaped and the submedian lobe is absent or variously developed. The cephalic framework is strongly sclerotized. The stylet is massive; the conus is much longer than the shaft and basal knobs. The basal knobs are anchor shaped. The isthmus is very short. The glandular pharyngeal basal bulb is greatly reduced.

In males, the stylet is absent. The spicules are of various shapes and sizes. The bursa is absent to well developed.

In juveniles, the cuticle has the same range of variation as in females but it may have different ornamentations in females and juveniles. When they are present, spines and scales are arranged in longitudinal rows; the stylet is not reduced but it is functional.

Females and juveniles are ectoparasites of plant roots, whereas males are nonfeeding and free-living in the soil.

Type genus

Criconema Hofmanner and Menzel, 1914

Key characteristics of some commonly found genera of Criconematidae

1. The body is almost straight; the annules in the anterior part of body are smooth, crenate, or serrated but scales, spines, and other outgrowths are absent. . . . . *Criconema*
2. Annules in the anterior part of body have scales, spines, and other outgrowths; medial body annules are lobed; a continuous fringe of spines is absent . . . .*Ogma*
3. The body is arcuate or ring-like; submedian lobes are separate; the vulva is not overhung by the anterior lip . . . . *Macroposthonia*
4. The first anterior annule is disc-like; the vulva is closed . . . . *Discocriconemella*
5. The first anterior annule is not disc-like; submedian lobes are present . . . . *Criconemoides*
6. The terior annule is not disc-like; submedian lobes are absent . . . . *Criconemella*

Genus recorded

Macroposthonia De Man, 1880.

Genus *Macroposthonia* De Man, 1880 (Figs. 3.72 and 3.73)

Diagnosis: The body is medium-sized and ventrally arcuate upon relaxation. The cuticle is thick; the annules have rounded, smooth, rough, or crenate margins posteriorly. The cephalic region is low, with two to three annules; the framework is strongly sclerotized; and submedian lobes are well developed. The stylet size and length are variable, from small with rounded knobs to large with anchor-shaped knobs. The vulva is near the anus. The postvulval uterine sac is absent. The tail is short, conical, or rounded. Males have two to four incisures each in the lateral fields and a reduced or distinct bursa.

Type species

Macroposthonia annulata De Man, 1880

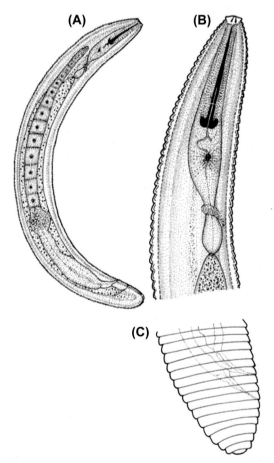

FIGURE 3.72   *Macroposthonia* sp.: (A) female; (B) pharyngeal region; (C) posterior region.

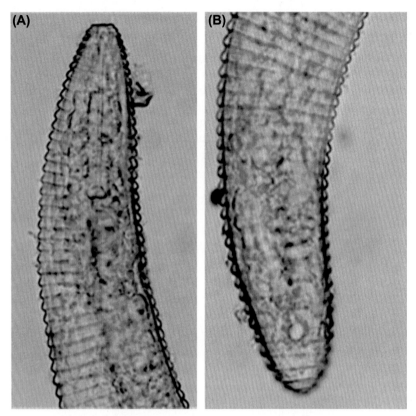

FIGURE 3.73 *Macroposthonia* sp.: (A) anterior end; (B) posterior region.

Commonly found species of *Macroposthonia* De Man, 1880

M. annulata De Man, 1880
M. curvata (Raski, 1952) De Grisse and Loof, 1965
M. microdorus (De Grisse, 1964) De Grisse and Loof, 1965
M. rustica (Micoletzky, 1915) De Grisse and Loof, 1965

Superfamily Tylenchuloidea Skarbilovich, 1947

Diagnosis: Marked sexual dimorphism is present. The stylet is degenerated or absent in males. The pharynx is always degenerated. The body is small, slender, swollen, or obese in females. The cuticle is generally thin with fine annulations, except in swollen or obese forms. The lateral fields have two to four incisures each, except in swollen or obese forms. The lip region is continuous with the body; smooth, submedian lobes are very weakly developed or absent. The stylet length is variable and usually delicate. The isthmus is slender. The basal bulb is small and rounded. The vulva is generally a transverse slit. The ovary is outstretched and coiled in obese or swollen females. The spicules are arcuate and setose with a fine tip. The bursa is absent, except in *Tylenchocriconema*. Juveniles are similar to females except for the absence of a stylet.

Type family

Tylenchulidae Skarbilovich, 1947

Other families

Paratylenchidae Thorne, 1949*
Sphaeronematidae Raski and Sher, 1952

Family Paratylenchidae Thorne, 1949

Diagnosis: These are usually vermiform, short, and plump nematodes. Sexual dimorphism is present in the anterior region. Males have a degenerated or no stylet and a degenerated pharynx. Adult females are vermiform. Body annulations are fine to moderate. The lateral fields have two to four incisures each. The stylet and pharynx are well developed in juveniles and females. The conus is longer than the shaft and basal knobs together. The basal knobs are small and rounded. The precorpus merges with the metacorpus; the isthmus is slender; and the basal bulb is small and offset. The vulva is a transverse slit located posteriorly; lateral flaps are either present or absent. The rectum and anus are sometimes indiscernible. Males are usually degenerated. The stylet is weakly developed or absent; the pharynx is obscure and occasionally degenerated. The spicules are cephalated, small, and slender. The bursa is rarely present.

Type genus

Paratylenchus Micoletzky, 1922.

Key characteristics of some commonly found genera of Paratylenchidae

1. The cephalic framework is conspicuous; the females are vermiform or slightly swollen at the vulval region but not obese or cylindroid; the postvulval region is more than one vulval body diameter (VBD) . . . . . *Paratylenchus*
2. The cephalic framework is obscure; the females are obese or cylindroid; the postvulval region is less than one VBD . . . . *Cacopaurus*
3. The cephalic framework is weak; the stylet length is 43–119 μm; the prevulval region is usually abnormally swollen . . . . *Gracilacus*

Genus recorded

Only the type genus was found in our collection.

Genus *Paratylenchus* Micoletzky, 1922 (Figs. 3.74 and 3.75)

Diagnosis: The body is short, usually less than 0.5 mm; it is vermiform, cylindrical, not abnormally swollen, curved ventrally, and usually C shaped upon relaxation. Body annulations are fine to moderate. The lateral fields have two to four

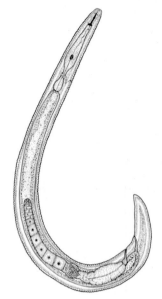

FIGURE 3.74 *Paratylenchus* sp.: female.

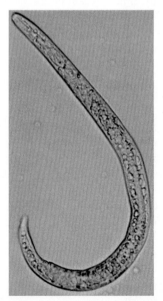

FIGURE 3.75 *Paratylenchus* sp.: female.

incisures each. The lip region is continuous with the body contour. The cephalic framework is weakly or strongly sclerotized. The stylet length is variable, from short to long (12–120 μm); the basal knobs are rounded. The precorpus and metacorpus are fused; and the valve plates are elongated. The isthmus is slender and the basal bulb is usually pyriform and sometimes rounded. The excretory pore is usually in the posterior region of the pharynx; sometimes it is anterior. The female genital system is monoprodelphic; the gonad is outstretched. The postvulval uterine sac is absent. The body behind the vulva is elongated. Males are slender; the stylet is degenerated or absent. The spicules are slender and arcuate. The gubernaculum is short. The bursa is absent.

Type species

Paratylenchus bukowinensis Micoletzky, 1922

Commonly found species of *Paratylenchus* Micoletzky, 1922

P. crenatus Corbett, 1966
P. goodeyi Oostenbrink, 1953
P. mutabilis Colbran, 1969
P. straeleni (De Coninck, 1931) Oostenbrink, 1960

Order: Aphelenchi Siddiqi, 1980

Diagnosis: These are small to medium-sized nematodes. The cuticle does not have prominent striations. The amphidial apertures are located laterosubdorsally. The stylet is small; basal knobs of the stylet are represented by thickening. The dorsal esophageal gland opens in the median bulb. The median bulb is strongly developed and has valve plates. The esophageal glands may or may not overlap the intestine in the form of a lobe. The vulva is located posteriorly (at more than 60% of the body length). Genital papillae are present in males. When they are large, bursa has bursal (papillary) ribs or small bursa at the tip of the tail. The spicules are mostly thorn shaped.

Type and only suborder

Aphelenchina Geraert, 1966.

Suborder: Aphelenchina Geraert, 1966

Diagnosis: Same as previously.

Type and only superfamily

Aphelenchoidea (Fuchs, 1937) Thorne, 1949.

Superfamily Aphelenchoidea (Fuchs, 1937) Thorne, 1949

Diagnosis: The cuticle has fine transverse striations. The lip region is continuous or is set off from the body by a constriction. The stylet has or does not have basal knobs. The median pharyngeal bulb is strong and muscular, and circular, ovoid, or squarish, with strong crescentic valve plates. The basal pharyngeal part forms an elongated pyriform bulb or is modified into a lobe extending over the intestine.

Type family

Aphelenchidae (Fuchs, 1937) Steiner, 1949*

Other families

Aphelenchoididae (Skarbilovich, 1947) Paramonov, 1953*
Paraphelenchidae (Goodey, 1951) Goodey, 1960
Anomictidae Goodey, 1960

Family Aphelenchidae (Fuchs, 1937) Steiner, 1949

Diagnosis: The stylet has no basal knobs. The pharyngeal glands may or may not overlap the intestine. The orifice of the dorsal pharyngeal gland nucleus usually opens in the median bulb. Males are with or without a bursa; if a bursa is present, it is supported by four bursal ribs; long, arcuate spicules have no flanges and a gubernaculum.

Type genus

Aphelenchus Bastian, 1865

Key to some commonly found genera of Aphelenchidae

1. The stylet has no basal knobs; the esophageal glands overlap the intestine dorsally; the female tail is short and bluntly rounded; the spicules are slender with a minute rostrum; and the bursa is well developed and supported by the bursal ribs . . . . *Aphelenchus*
2. The stylet may or may not have basal knobs; the pharyngeal glands do not form a lobe; the excretory pore is posterior to the median bulb; the female tail is short and conical, with or without a terminal mucron; the male tail is conoid and often mucronate or digitate; there are three to five pairs of genital papillae; and the bursa are either present or absent . . . . *Paraphelenchus*

Genus recorded

Only the type genus was found in our collection.

Genus *Aphelenchus* Bastian, 1865 (Figs. 3.76 and 3.77)

Diagnosis: The body tapers at both ends, more sharply anteriorly. The body cuticle has transverse striations. The cephalic cuticle is smooth. The lateral fields each have about 10 incisures. Deirids are present. The lip region is slightly set off from the body. The stylet is moderately developed; basal knobs are absent. The procorpus is more or less cylindrical and constricted at the junction with

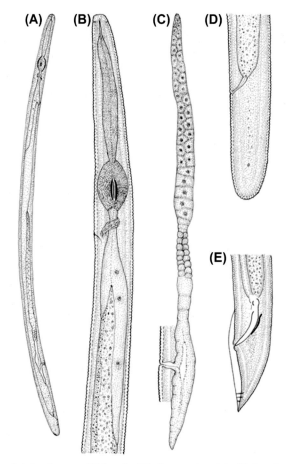

FIGURE 3.76 *Aphelenchus* sp.: (A) female; (B) pharyngeal region; (C) female genital system; (D) female posterior region; (E) male posterior region.

the median bulb. The median bulb is strongly muscular, with crescentic valve plates. The pharyngeal glands are in a lobe extending over the intestine dorsally. The excretory pore is in the isthmus region. The females have a posteriorly located vulva and a monoprodelphic genital system. The postvulval uterine sac is short. The tail is short and cylindrical with a rounded terminus. Males have paired, slender, arcuate spicules and a gubernaculum. The bursa is well developed, with four pairs of bursal papillae or ribs.

Type species

Aphelenchus avenae Bastian, 1865

Commonly found species of *Aphelenchus* Bastian, 1865

A. avenae Bastian, 1865

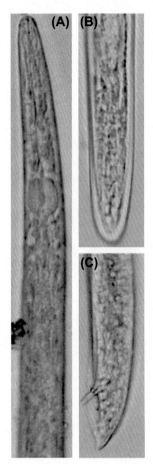

FIGURE 3.77   *Aphelenchus* sp.: (A) pharyngeal region; (B) female posterior region; (C) male posterior region.

Genus *Paraphelenchus* (Micoletzky, 1922) Micoletzky, 1925
(Figs. 3.78 and 3.79)

Diagnosis: The body is usually small and tapers slightly anteriorly. The cuticle has transverse striations. The lip region is continuous with the body contour. The basal pharyngeal bulb is not modified into a lobe; glands are located in a bulb. The excretory pore is in the isthmus region. Females have a monoprodelphic genital system. The vulva is located posteriorly. A postvulval uterine sac is present. The tail is bluntly conoid, sometimes with a terminal mucro. Males have slender spicules, a rodlike gubernaculum, and five pairs of genital papillae. The first pair of genital papillae is present, and is represented by a single adcloacal papilla, or is absent.

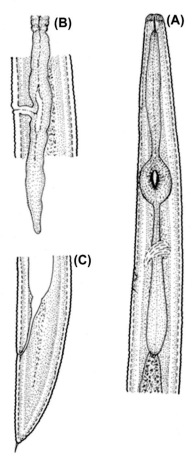

FIGURE 3.78 *Paraphelenchus* sp.: (A) pharyngeal region; (B) vulval region; (C) female posterior region.

Type species

Paraphelenchus pseudoparietinus (Micoletzky, 1922) Micoletzky, 1925

Commonly found species of *Paraphelenchus* (Micoletzky, 1922) Micoletzky, 1925

P. *acontioides* Taylor and Pillai, 1967
P. *myceliophthorus* Goodey, 1958.

Family Aphelenchoididae (Skarbilovich, 1947) Paramonov, 1953

Diagnosis: Pharyngeal glands are present in a lobe overlapping the intestine dorsally. Spicules are usually shaped like the thorn of a rose. The gubernaculum

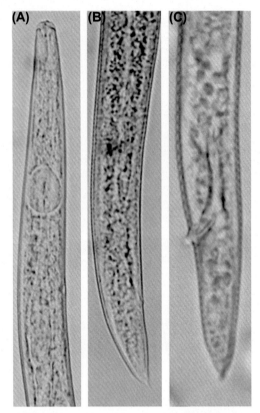

FIGURE 3.79   *Paraphelenchus* sp.: (A) pharyngeal region; (B) female posterior region; (C) male posterior region.

is absent. The terminal bursa in the tail of males is usually absent. Rarely, it is present, or sometimes it is represented by thickening in the form of bursal folds. There are one to three pairs of genital papillae.

Type genus

Aphelenchoides Fischer, 1894

Key to some commonly found genera of Aphelenchoididae

1. The stylet length is less than 30 pm; V is less than 80%; the bursa is present; the gubernaculum is absent; the spicules are not fused; the anus is present; the tail is strongly curved, without an appendage . . . .*Rhadinaphelenchus*
2. The stylet has small, rounded basal knobs; the vulval lips sometimes protrude; the female tail is rounded, conoid, or sharply pointed; the spicules usually have a prominent rostrum; the male tail has terminal

bursa; there are two pairs of genital papillae: one adanal and one postanal; the gubernaculum is absent . . . .*Bursaphelenchus*

3. The stylet has basal knobs; the spicules are shaped liked the thorns of a rose; the bursa and gubernaculum are absent; there are three pairs of genital papillae; the tails of both sexes more or less taper, frequently with one or more mucrons . . . .*Aphelenchoides*

4. The body is long and slender; the stylet has small basal knobs; the vulva is at 85%–90% of the body, mostly narrowing suddenly posterior to the vulva; the male tail is short and conical, with a short terminal bursa; there are three pairs of genital papillae; the gubernaculum is absent; juveniles have cuticular projections at both ends . . . . *Parasitaphelenchus*

5. The stylet has small basal knobs; the median bulb is pear shaped; the isthmus is absent; a postvulval uterine sac is present; the tail is ventrally strongly curved distally; the bursa is supported by bursal rib-like genital papillae . . . . *Pseudaphelenchus*

Genera recorded

Aphelenchoides Fischer, 1894
Pseudaphelenchus Kanzaki et al., 2009
Bursaphelenchus Fuchs, 1937

Genus *Aphelenchoides* Fischer, 1894 (Figs. 3.80 and 3.81)

Diagnosis: The body is usually long and slender, and narrows slightly behind the vulva, further behind the anal region. The cuticle has fine, transverse striations. The lateral fields each have a few incisures. The lip region is set off from the body. The stylet either does or does not have basal knobs. The procorpus is usually cylindrical and is constricted at the junction with the median bulb. The median bulb is strongly muscular, with crescentic valve plates. The pharyngeal glands are in a lobe extending over the intestine dorsally. The females have a posteriorly located vulva and a monoprodelphic genital system. The postvulval uterine sac is short or sometimes it is absent. The tail is short and more or less tapers to a conoid terminus. The tail terminus usually has one or more mucrons. Males have paired, rose thorn-like, arcuate spicules. The gubernaculum and bursa are absent. There are often three pairs of genital papillae.

Type species

Aphelenchoides kuehnii Fischer, 1894

Commonly found species of *Aphelenchoides* Fischer, 1894

A. besseyi Christie, 1942
A. fragariae (Ritzema Bos, 1891) Christie, 1932
A. ritzemabosi (Schwartz, 1911) Steiner and Buhrer, 1932

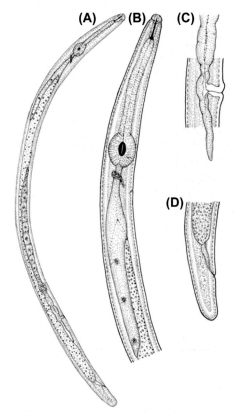

FIGURE 3.80 *Aphelenchoides* sp.: (A) female; (B) pharyngeal region; (C) vulval region; (D) female posterior region.

Genus *Pseudaphelenchus* Kanzaki et al., 2009 (Figs. 3.82 and 3.83)

Diagnosis: The body is small to medium-sized and is ventrally arcuate or straight upon fixation. The cuticle is thin and annulated. The cephalic region is set off from the body; the lip region is truncated. The stylet has small but conspicuous basal knobs. The procorpus is cylindrical; the median bulb is pear shaped. The postcorpus is glandular and extends over the intestine dorsally. The female genital system is monoprodelphic; a postvulval uterine sac is present. The vagina is straight; the vulva is a simple slit. The tail strongly curves ventrally distally.

Type species

Pseudaphelenchus yukiae Kanzaki et al., 2009
Pseudaphelenchus is not a common genus. It is represented by only a few species and so far it has been reported in a limited number of places.

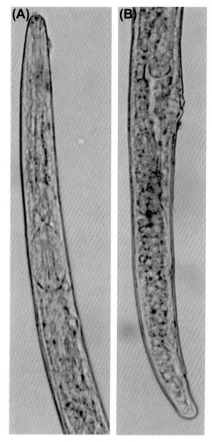

FIGURE 3.81 *Aphelenchoides* sp.: (A) pharyngeal region; (B) vulval and posterior region.

Genus *Bursaphelenchus* Fuchs, 1937 (Figs. 3.84 and 3.85)

Diagnosis: The body is vermiform and 0.3–1.7 mm in length. The cuticle has fine annules. The lip region has no oral disc; the lateral lips are smaller than the subdorsal and subventral ones. The stylet is slender and the lumen narrow, with a length less than 30 μm; the basal knobs are weakly developed. The median pharyngeal bulb is well developed. The pharyngeal glands extend over the intestine as a lobe. The anterior vulval flap is either present or absent. The postvulval uterine sac is three to six vulval body widths long. Males have a strongly ventrally curved posterior region. The spicules are not fused; they are hook-like and rarely linear. They are never strongly curved and usually have a prominent rostrum. The gubernaculum is absent. There are two or more pairs of genital papillae: one is adanal and one to four pairs are postcloacal. The tail is subconoid and

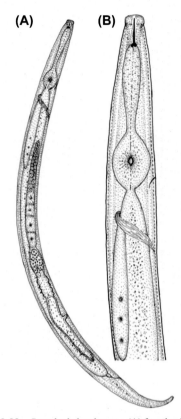

(A)   **(B)**

FIGURE 3.82   *Pseudaphelenchus* sp.: (A) female; (B) pharynx.

evenly tapered. The tip is usually smooth, sometimes with a simple mucro, but not spicate or with four tubercles. The bursa is represented by a flap-like cuticle.

Type species

Bursaphelenchus piniperdae Fuchs, 1937

Commonly found species of *Bursaphelenchus* Fuchs, 1937

B. cocophilus (Cobb, 1919) Baujard, 1989
B. mucronatus Mamiya and Enda, 1979
B. xylophilus (Steiner and Buhrer, 1934) Nickle, 1970

Order Dorylaimida Pearse, 1942

Diagnosis: The cuticle is smooth or has fine or coarse striations. There are six lips, each provided with three labial papillae (one in the inner margin and two in the outer margin) except the laterals (one in the inner and one in the outer). Amphids are variously shaped (cyathiform, pouch-like, or tubular) with

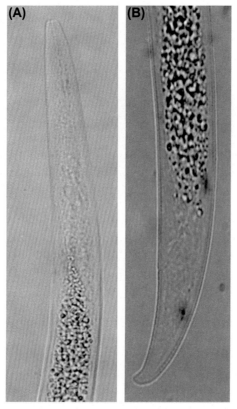

FIGURE 3.83   *Pseudaphelenchus* sp.: (A) pharynx; (B) female posterior region.

pore- or slit-like apertures. The feeding apparatus is composed of a vestibulum, a guiding ring(s)/apparatus, and an odontostyle/mural tooth. The vestibulum is variable; it is weakly or heavily sclerotized and may or may not be provided with denticles or onchia. The pharynx is divisible into two parts: an anterior slender part and a posterior expanded one. The posterior expanded part houses five or, rarely, three pharyngeal glands. Orifices of the pharyngeal gland nuclei are in the expanded part itself. The nerve ring encircles the anterior slender part. The excretory pore is absent. The pharyngo-intestinal junction has a well-developed cardia. The female genital system is amphidelphic, monoprodelphic, or mono-opisthodelphic, always with reflexed ovary/ies. The vulva is a transverse or longitudinal slit or is pore-like. Males have a pair of testes. The spicules are free and usually robust; the lateral guiding pieces are present; the gubernaculum is absent or very rarely present. A few to numerous ventromedian supplements are present or, rarely, absent. A prerectum is present. The tail is variable in shape and length, and is similar or dissimilar in the sexes.

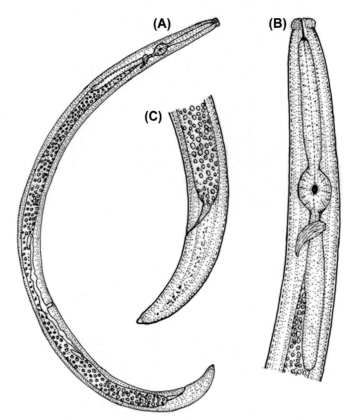

FIGURE 3.84 *Bursaphelenchus* sp.: (A) female; (B) pharyngeal region; (C) female posterior region.

Type suborder

Dorylaimina Pearse, 1936*

Other suborders

Nygolaimina Ahmad and Jairajpuri, 1979*
Campydorina Jairajpuri, 1983*

The systematic position of Campydorina is doubtful; it has been placed under Enoplida/Ironida, etc. There is no doubt that it is not a nematode of the order Dorylaimida.

Suborder Dorylaimina Pearse, 1936

Diagnosis: Amphids are usually cyathiform or pouch-shaped. The feeding apparatus consists of an anterior odontostyle and a posterior odontophore. The odontostyle moves through a single or double guiding ring(s). Odontophores

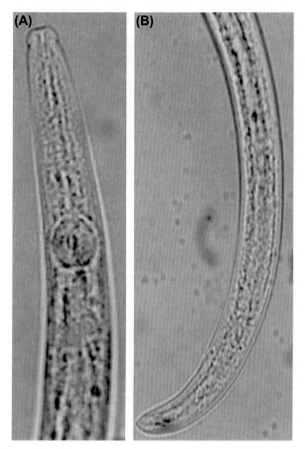

FIGURE 3.85   *Bursaphelenchus* sp.: (A) pharyngeal region; (B) female posterior region.

are variously shaped: they may be simply rodlike or have basal knobs or basal flanges. The anterior slender part of the pharynx is shorter than or equal to the expanded part, or the expanded part may be limited to a small pyriform bulb. Males have an adcloacal pair and a series of ventromedian supplements; these are rarely absent. The ventromedians may be spaced apart, contiguous, or grouped. The prerectum is usually distinguishable and varies in length.

Type superfamily

Dorylaimoidea De Man, 1876*

Other superfamilies

Tylencholaimoidea Filipjev, 1934*
Longidoroidea Thorne, 1935*
Actinolaimoidea Thorne, 1939
Belondiroidea Thorne, 1939*

Superfamily Dorylaimoidea De Man, 1876

Diagnosis: These nematodes are generally medium-sized to large. The amphids have stirrup-shaped or pouch-like fovea and slit-like apertures. The odontostyle is a cylindrical opening/aperture that is dorsally placed and obliquely seen in the lateral view. The odontophore is usually rodlike and sometimes is provided with basal knobs or flanges. The pharynx has slender anterior and expanded posterior parts. The females have monodelphic or amphidelphic gonads. The vulva is a transverse or longitudinal slit or is simply pore-like. The males have dorylaimoid spicules with lateral guiding pieces and a number of ventromedian supplements that may be spaced, contiguous, or grouped.

Type family

Dorylaimidae De Man, 1876*

Other families

Nordiidae Jairajpuri and Siddiqi, 1964*
Aporcelaimidae Heyns, 1965*
Qudsianematidae Jairajpuri, 1965*

Family Dorylaimidae De Man, 1876

Diagnosis: The body cuticle has fine transverse striations. The amphidial apertures are slit-like. The odontostyle is one to one-and-a-half lip region widths long, with an aperture one-third its length. The odontophore is rodlike; basal knobs and flanges are absent. The guiding ring is either single or double. The basal expanded part of the pharynx is less than or equal to the length of the anterior slender part. The females have amphidelphic or, rarely, mono-opisthodelphic gonads. Lateral guiding pieces are present. The gubernaculum is absent. The tail is elongated to long and filiform in the female. It is either similar or dissimilar between the sexes.

Type genus

Dorylaimus Dujardin, 1845

Key characteristics of some commonly found genera of Dorylaimidae

1. The tail is long and filiform in both sexes 2-3
   Tail is elongated to long and filiform in the female and short and conoid in the male. 4-9
2. The lip region is usually continuous; the spicules are dorylaimoid; there are numerous ventromedian supplements . . . . *Prodorylaimus*
3. The lip region is offset; the spicules are nondorylaimoid; there is only one ventromedian supplement . . . . *Amphidorylaimus*
4. The cuticle has longitudinal ridges; the body is large; the odontostyle is about 1.5 lip regions wide; there are about contiguous 25–55 ventromedian supplements; . . . . *Dorylaimus*

5. The cuticle has longitudinal ridges; the body is large; the odontostyle is about 1.5 lip regions wide; the ventromedian is supplemented by only two to three groups (fascicles)....................... *Ischiodorylaimus*
6. The cuticle has no longitudinal ridges; the odontostyle is about one lip region wide... 7-9
7. The double contour on the dorsal side of the spicules and the spur near the tip are absent; the supplements are well developed..... *Mesodorylaimus*
8. The spicules are strong; a double contour on the dorsal side and a spur near the tip are present; supplements are very small..................
   .... *Calcaridorylaimus*
9. The labial framework is sclerotized; the odontostyle is narrow.........
   .... *Thornenema*

Genera recorded

Mesodorylaimus Andrássy, 1959
Prodorylaimus Andrássy, 1959

Genus *Mesodorylaimus* Andrássy, 1959 (Figs. 3.86 and 3.87)

Diagnosis: These are medium-sized nematodes, usually 1–2 mm in length. The cuticle is thick and devoid of longitudinal striations; transverse striations are fine and not clear under the light microscope. The lip region is continuous or offset; the lips are either separate or fused and angular or rounded. The odontostyle is almost as long as the lip region width; the aperture is about one-third its length. The odontophore is simple and rodlike, and not modified. The pharyngo-intestinal junction has a small, conoid cardia. The vulva is a transverse slit that is sometimes longitudinal; vulval sclerotization is strong. Males have a short prerectum reaching only the ventromedian supplements or slightly more. The spicules are strong and dorylaimoid; lateral guiding pieces are present. The number of ventromedian supplements varies from 4 to 26; they are usually adjacent or spaced. The female tail is elongated to long and filiform; the male tail is short and hemispheroid.

Type species

Mesodorylaimus mesonyctius (Kreis, 1930) Andrássy, 1959

Commonly found species of *Mesodorylaimus* Andrássy, 1959

M. bastiani (Bütschli, 1873) Andrássy, 1959
M. litorallis Loof, 1969
M. mesonyctius (Kreis, 1930) Andrássy, 1959

Genus *Prodorylaimus* Andrássy, 1959 (Figs. 3.88 and 3.89)

Diagnosis: The body size varies from 1 to 5 mm. The cuticle is smooth or has fine striations. The lip region is offset or is continuous with the body; the labial papillae are usually raised. The amphids have stirrup-shaped fovea and slit-like apertures. The odontostyle is strong, with a wide lumen. The odontophore is

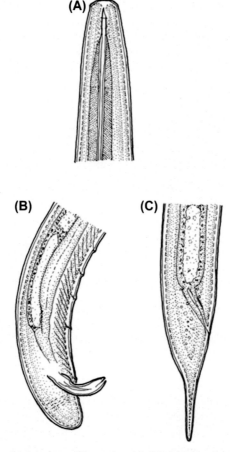

FIGURE 3.86 *Mesodorylaimus* sp.: (A) anterior end; (B) male posterior region; (C) female posterior region.

simple and rodlike. The females have a thick-walled vagina and an amphidelphic genital system. The males have arcuate spicules, lateral guiding pieces, and numerous contiguous ventromedian supplements. The ventromedian supplements start well above the range of spicules when they are retracted. The tails in both sexes are elongated and conoid to long and filiform.

Type species

Prodorylaimus longicaudatus (Butschli, 1874) Andrássy, 1959

Commonly found species of *Prodorylaimus* Andrássy, 1959

P. filiarum Andrássy, 1964
P. longicaudatus (Butschli, 1874) Andrássy, 1959
P. mas Loof, 1985
P. paraagilis (Altherr, 1953) Andrássy, 1986

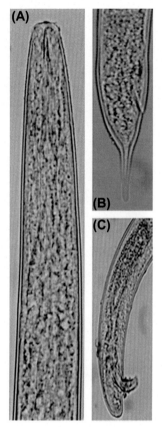

FIGURE 3.87 *Mesodorylaimus* sp.: (A) anterior end; (B) female posterior region; (C) male posterior region.

Family Aporcelaimidae Heyns, 1965

Diagnosis: These are generally large nematodes except a few medium-sized species. These are ventrally arcuate upon relaxation. The cuticle is usually thick with fine transverse striations and crisscross lines or punctations. Paired, irregularly spaced body pores in the lateral sector are generally present. The lip region is distinctly set off; lips are separated. The odontostyle has a wide lumen and aperture or a mural tooth. The guiding ring is thin, plicated, and flap-like. The odontophore is simple and rodlike. The basal expanded part of the pharynx has irregular tubules. A thin epithelial sheath is present covering the expanded part. The pharyngo-intestinal junction has a well-developed cardia, sometimes with a flattened disc-shaped structure or glands. The females have amphidelphic gonads. The spicules are dorylaimoid, with lateral guiding pieces present. The ventromedian supplements are spaced. The tail is short and conoid to hemispheroid; it is rarely long and filiform, and is similar in both sexes.

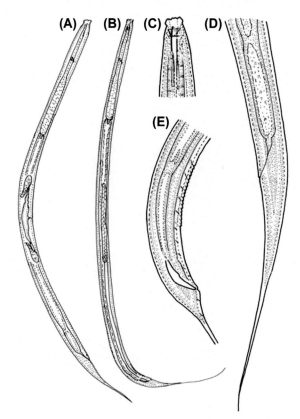

FIGURE 3.88 *Prodorylaimus* sp.: (A) female; (B) male; (C) anterior end; (D) female posterior region; (E) male posterior region.

Type genus

Aporcelaimus Thorne and Swanger, 1936

Key characteristics of some commonly found genera of Aporcelaimidae

1. The feeding apparatus is modified to become a mural tooth . . *Sectonema*
   The feeding apparatus is an axial spar (odontostyle) 2
2. Layering of cuticle is inconspicuous; crisscross lines or punctation are absent 3, 4
   Layering of the cuticle is conspicuous, with transverse markings or crisscross lines or punctations 5-8
3. The females have a long, filiform tail . . . . *Aporcedorus*
4. The females have a short, conoid tail. . . . . *Aporcelaimium*
5. The body is large; the pharynx has no cardiac glands, with faint longitudinal markings or crisscross lines present on the cuticle . . . . *Aporcelaimus*

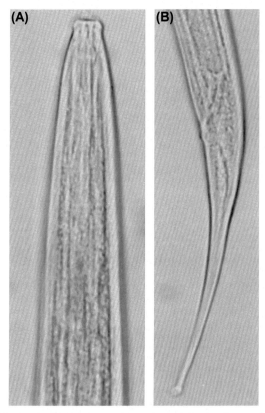

FIGURE 3.89 *Prodorylaimus* sp.: (A) anterior end; (B) female posterior region.

6. The lip region is offset by a slight depression; the vulva is longitudinal
.... *Makatinus*
7. The cuticle has transverse striations; the oral aperture is hexagonal
.... *Aporcelaimellus*
8. The cuticle is thick; the odontostyle is small, wide, and slightly bent;
the pharyngo-intestinal junction has well-developed cardiac glands
.... *Paraxonchium*

Genera recorded

Aporcelaimus Thorne and Swanger, 1936
Aporcelaimellus Heyns, 1965
Makatinus Heyns, 1965
Sectonema Thorne, 1930

Genus *Aporcelaimus* Thorne and Swanger, 1936 (Figs. 3.90 and 3.91)

Diagnosis: The body length is variable and generally large, usually not less than 4 mm. The cuticle is thick, with interwoven striations, and rarely with

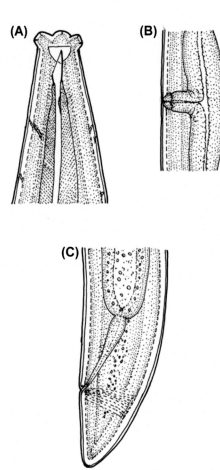

FIGURE 3.90 *Aporcelaimus* sp.: (A) anterior end; (B) vulval region; (C) female posterior region.

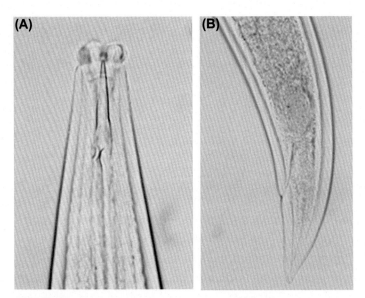

FIGURE 3.91 *Aporcelaimus* sp.: (A) anterior end; (B) female posterior region.

fine transverse striations. The amphids are stirrup shaped with duplex fovea; some species have apparent, sclerotized median support. Apertures are slit-like. The lip region is set off from the body by a constriction. The lips are separated. The odontostyle is axial and short with a wide lumen. The aperture is more than half its length. The odontophore is rodlike. Narrow and expanded parts of the pharynx are of almost a similar length. The cardiac disc is conspicuous. The females have amphidelphic gonads. The vulva is sclerotized; the aperture is a large transverse slit. The spicules are strong and arcuate. The ventromedian supplements vary from a few to many and spaced. The tails in both sexes are similar: blunt, rounded, convex-conoid, or hemispherical.

Type species

Aporcelaimus regius (De Mana, 1876) Thorne and Swanger, 1936

Commonly found species of *Aporcelaimus* Thorne and Swanger, 1936

A. regius (De Mana, 1876) Thorne and Swanger, 1936
A. sublabiatus (Thorne and Swanger, 1936) Brzeski, 1962
A. superbus (De Man, 1880) T. Goodey, 1951

Genus *Aporcelaimellus* Heyns, 1965 (Figs. 3.92 and 3.93)

Diagnosis: These are usually medium-sized nematodes. The cuticle is thick and three-layered, more distinct in the caudal region. A hyaline space is present between the intermediate layers at the tail terminus. Dorsal and ventral body pores are generally distinct in the anterior region. The amphids are stirrup shaped with duplex fovea; some species have apparent sclerotized median support; apertures are slit-like. The lip region is set off from the body by a constriction; the lips are separated and angular. The oral aperture is hexagonal. The odontostyle is axial, short, and well built with a wide lumen; the aperture is almost half its length. The odontophore is simple and rodlike. A cardiac disc is usually present. The females' genital system is amphidelphic. The vulva is a transverse slit or pore-like. Spicules are strong and arcuate, with lateral guiding pieces present; there are seven to 25 ventromedian supplements that are not contiguous. The tail in both sexes is similar: short, rounded, conical, or sometimes digitated.

Type species

Aporcelaimellus obscurus (Thorne and Swanger, 1936) Heyns, 1965

Commonly found species of *Aporcelaimellus* Heyns, 1965

A. obscurus (Thorne and Swanger, 1936) Heyns, 1965
A. paraobtusicaudatus (Micoletzky, 1922) Andrássy, 1986
A. simplex (Thorne and Swanger, 1936) Loof and Coomans, 1970

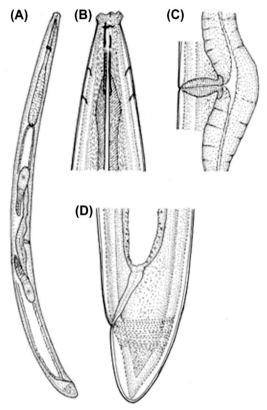

FIGURE 3.92 *Aporcelaimellus* sp.: (A) female; (B) anterior end; (C) vulval region; (D) female posterior region.

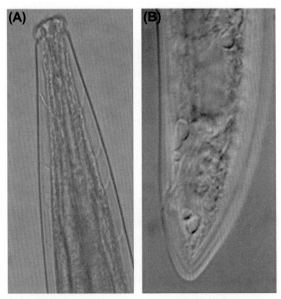

FIGURE 3.93 *Aporcelaimellus* sp.: (A) anterior end; (B) female posterior region. *Image (B) was photographed by Wasim Ahmad.*

Genus *Makatinus* Heyns, 1965 (Figs. 3.94 and 3.95)

Diagnosis: The body is usually large, 3–4mm in length. The cuticle is superficially punctated. The lip region is slightly offset from the body; the lips are compact. The odontostyle is dorylaimoid, with a wide lumen and aperture. The odontophore is simple and rodlike. The anterior part of the pharynx is cruciform. The pharyngo-intestinal junction has no cardiac disc or glands. The females have a cuticularized, longitudinal vulva and an amphidelphic genital system. The males have dorylaimoid spicules and numerous ventromedian supplements; adcloacal supplements are in two to three pairs. The tail in both sexes is short and conoid.

Type species

Makatinus punctatus Heyns, 1965

Commonly found species of *Makatinus* Heyns, 1965

M. macropunctatus Heyns, 1967
M. heynsi Ahmad and Ahmad, 1992

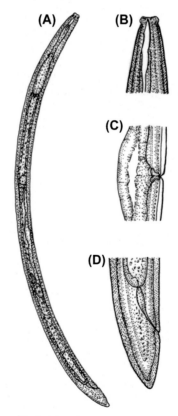

FIGURE 3.94   *Makatinus* sp.: (A) female; (B) anterior end; (C) vulval region; (D) posterior region.

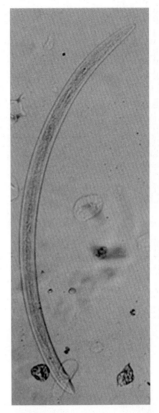

FIGURE 3.95 *Makatinus* sp.: male.

Genus *Sectonema* Thorne, 1930 (Figs. 3.96 and 3.97)

Diagnosis: The body is large, usually more than 4 mm in length. The cuticle has transverse and interwoven striations, rarely only with transverse striations. The amphids have sclerotized median support and duplex fovea. The cephalic region is set off from the body; the lips are separated. The stoma has a mural tooth attached to the subventral wall. The mural tooth has a dorsal groove directed anteriorly. The pharyngo-intestinal junction has a conoid cardia; a cardiac disc is present. The females have an amphidelphic genital system. The vulva is transverse; vulval sclerotization is present. The spicules are dorylaimoid and well built; ventromedian supplements are spaced and irregularly arranged. The tail in both sexes is short and conoid to hemispheroid.

Type species

Sectonema ventrale Thorne, 1930

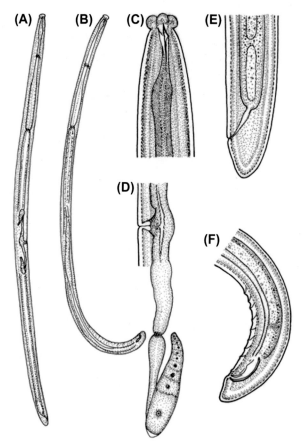

FIGURE 3.96 *Sectonema* sp.: (A) female; (B) male; (C) anterior end; (D) female genital system (posterior); (E) female posterior region; (F) male posterior region.

Commonly found species of *Sectonema* Thorne, 1930

S. barbatoides Heyns, 1965
S. demani Altherr, 1965
S. rotundicauda Goodey, 1951

Family Qudsianematidae Jairajpuri, 1965

Diagnosis: The cuticle is smooth or has fine striations. The cephalic region is continuous with the body or is offset by a constriction; the lips are variable. The stoma has a cylindrical odontostyle with a clear lumen and a wide aperture. The odontophore is simple and rodlike or is modified with basal flanges or knobs. The pharynx is muscular. The pharyngo-intestinal junction has a hemispheroid or conoid

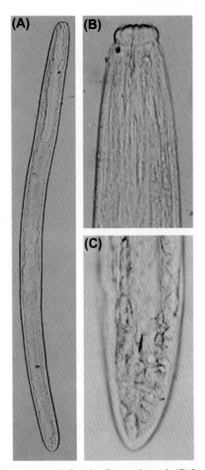

FIGURE 3.97 *Sectonema* sp.: (A) female; (B) anterior end; (C) female posterior region.

cardia; a cardiac disc is sometimes present. The females have mono-opisthodelphic or amphidelphic gonads. The spicules are dorylaimoid; lateral guiding pieces are usually present; they are rarely absent. Supplements are variable in number (a few to many) and arrangement (spaced or contiguous). The tails in both sexes are similar: hemispheroid or short and conoid to elongated and conoid.

Type genus

Qudsianema Jairajpuri, 1965

Key characteristics of some commonly found genera of Qudsianematidae

1. The amphids are labial 3, 4
2. The amphids are postlabial 5-11
3. The female genital system is mono-opisthodelphic . . . . *Ecumenicus*

4. The female genital system is amphidelphic. . . . . *Kochinema*
5. The odontostyle has a broad lumen; the circumoral area is not deeply sunken; the inner liplets are not well separated. . . . . *Labronema*
6. The lips are distinct; a cardiac disc is present; ventromedian supplements are spaced; the tails in both sexes are hemispheroid, convex and conoid, or rounded. . . . . *Crassolabium*
7. The odontostyle is thin walled; the odontophore is flanged; the expanded part of the pharynx has a constriction in middle . . . . *Qudsianema*
8. The lip region is discoid without cuticularized plates. . . . . *Discolaimus*
9. The lip region is not discoid and is set off by an expansion . . . .
. . . . *Discolaimoides*
10. The lips are predominantly angular; the pharynx is short; a precloacal space is present. . . . . *Eudorylaimus*
11. The cardiac disc is absent; the tail is elongated and conoid, 3.5–8.0 of the width of the anal body . . . .*Epidorylaimus*

Genera recorded

Ecumenicus Thorne, 1974
Epidorylaimus Andrássy, 1986
Eudorylaimus Andrássy, 1959
Discolaimus Cobb, 1913
Kochinema Heyns, 1963
Crassolabium Yeates, 1967

Genus *Ecumenicus* Thorne, 1974 (Figs. 3.98 and 3.99)

Diagnosis: These are small nematodes, usually less than 1.5 mm. The cuticle has fine transverse striations. The lateral sector of the body has a column of cells with a small, pore-like opening on each cuticle. The cephalic region is set off from the body by a shallow constriction; the lips are angular. The odontostyle has a wide lumen and aperture. The odontophore is rodlike and devoid of flanges or knobs. The narrow part of the pharynx is longer than the expanded part. The females have a mono-opisthodelphic gonad. The vulva has light sclerotization. The vagina is tubular and is directed posteriorly. The males lack lateral guiding pieces. The spicules are arcuate and well developed. The ventromedian supplements are spaced, usually eight in number. The tail terminus in both sexes is digitated.

Type species

Ecumenicus monohystera (De Man, 1880) Thorne, 1974

Genus *Epidorylaimus* Andrássy, 1986 (Figs. 3.100 and 3.101)

Diagnosis: The body is small to medium-sized. The cuticle may or may not have fine striations; striations are more prominent on the tail. The lip region

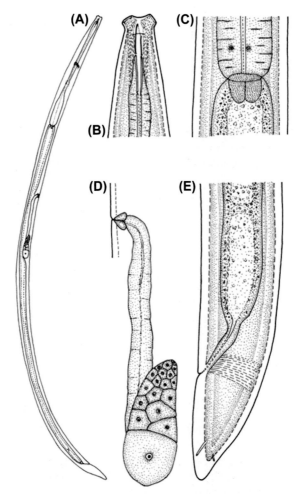

FIGURE 3.98 *Ecumenicus* sp.: (A) female; (B) anterior end; (C) esophago-intestinal junction; (D) female genital system; (E) posterior region.

is set off from the body; lips are angular. The odontostyle is one and a half times the length of the lip region. The guiding ring is single. The expanded part of the esophagus is half the pharyngeal length or less. The females have an amphidelphic reproductive system. Males are usually rare; if present, they have arcuate spicules, lateral guiding pieces, and four to nine ventromedian supplements. A precloacal space between ventromedian supplements and an adanal pair are absent. The tails in both sexes are elongated and conoid, and are ventrally curved with a pointed terminus.

Type species

Epidorylaimus lugdunensis (De Man, 1880) Andrássy, 1986

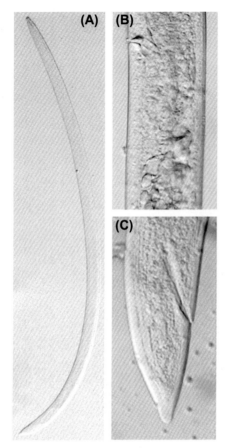

FIGURE 3.99 *Ecumenicus* sp.: (A) female; (B) vulval region; (C) posterior region.

Commonly found species of *Epidorylaimus* Andrássy, 1986

E. agilis (De Man, 1880) Andrássy, 1986
E. consobrinus (De Man, 1918) Andrássy, 1986
E. humilis (Thorne and Swanger, 1936) Andrássy, 1986

Genus *Eudorylaimus* Andrássy, 1959 (Figs. 3.102 and 3.103)

Diagnosis: The body is small to medium in size, ranging from 0.9 to 3.5 mm. The cuticle is usually smooth, with fine transverse striations present or absent. The amphidial fovea is stirrup shaped and the aperture is slit-like. The lip region is set off from the body contour; the lips are separated and mostly angular in outline. The odontostyle is about one lip region wide and the aperture is wide; the odontophore is simple and devoid of basal flanges or knobs. The expanded part of the pharynx is about half the neck length. The females have a transverse or, rarely,

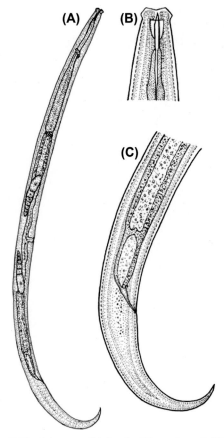

FIGURE 3.100   *Epidorylaimus* sp.: (A) female; (B) anterior end; (C) posterior region.

FIGURE 3.101   *Epidorylaimus* sp.

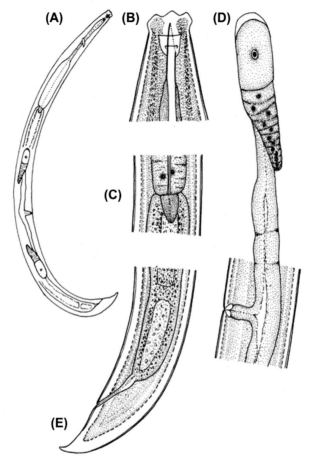

FIGURE 3.102 *Eudorylamus* sp.: (A) female; (B) anterior end; (C) pharyngo-intestinal junction; (D) female genital system; (E) posterior region.

a longitudinal vulva, and an amphidelphic genital system with reflexed ovaries. The prerectum is short to long. The spicules are well built and arcuate; lateral guiding pieces are present. There are 3 to 18 ventromedian supplements that are spaced, with a precloacal space between the adcloacal pair and the ventromedians. The tails in both sexes are short and conoid, or bluntly rounded; the terminus is generally ventrally curved; it is rarely straight or slightly dorsally bent.

Type species

Eudorylamus carteri (Bastian, 1865) Andrássy, 1959

Commonly found species of *Eudorylaimus* Andrássy, 1959

E. acuticauda (De Man, 1880) Andrássy, 1959
E. carteri (Bastian, 1865) Andrássy, 1959
E. centrocercus (De Man, 1880) Andrássy, 1959

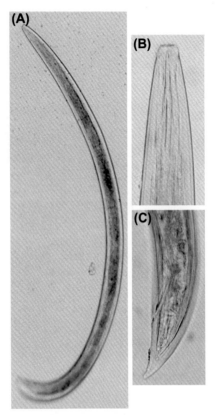

FIGURE 3.103   *Eudorylamus* sp.: (A) female; (B) anterior end; (C) posterior region.

E. leuckarti (Butschli, 1873) Andrássy, 1959
E. similis (De Man, 1876) Andrássy, 1959

Genus *Discolaimus* Cobb, 1913 (Figs. 3.104 and 3.105)

Diagnosis: These nematodes are small to medium-sized, generally 1–3 mm in length. The cuticle has fine transverse striations. The numerous, large, glandular bodies each have a single pore-like opening on the cuticle present in the lateral chords. The amphids are stirrup shaped with a slit-like aperture. The lip region is expanded, discoid, or sucker-like; the oral aperture is sunken. The odontostyle is well developed, usually with thick walls and a wide lumen. The odontophore is simple and rodlike. The anterior slender part of the pharynx is not uniform; it has no strong musculature and suddenly expands to form a strongly muscular basal part. The females have an amphidelphic or, very rarely, a mono-opisthodelphic genital system. The vulva is transverse; the vagina is not sclerotized. The males have dorylaimoid spicules, with lateral guiding pieces and supplements with an adcloacal pair and a series of spaced ventromedians. The tails in both sexes are short and rounded or bluntly conoid.

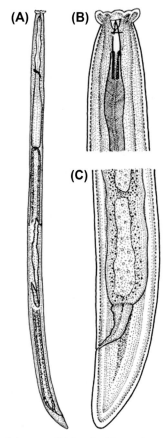

FIGURE 3.104 *Discolaimus* sp.: (A) female; (B) anterior end; (C) posterior region.

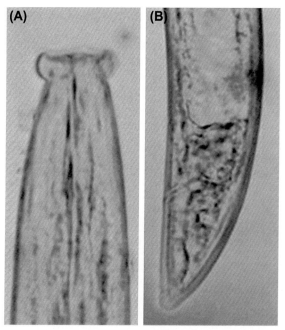

FIGURE 3.105 *Discolaimus* sp.: (A) anterior end; (B) posterior region.

Type species

Discolaimus texanus Cobb, 1913

Commonly found species of *Discolaimus* Cobb, 1913

D. major Thorne, 1939
D. similis Thorne, 1939
D. texanus Cobb, 1913

Genus *Kochinema* Heyns, 1963 (Figs. 3.106 and 3.107)

Diagnosis: The body is small to medium-sized. The lips are expanded but not discoid. The amphids have a slit-like aperture and are labial in position. The odontostyle is slender and long, and the aperture is small. The odontophore is rodlike. The pharynx has its one half expanded. The vagina is perpendicular to the body; the vulva is transverse. The females have amphidelphic gonads. The males have doylaimoid spicules, lateral guiding pieces, and a few spaced ventromedian supplements. The tails in both sexes are short and conoid.

Type species

Kochinema proamphidium Heyns, 1963

Commonly found species of *Kochinema* Heyns, 1963

K. proamphidium Heyns, 1963
K. longum Argo and van den Berg, 1971

Genus *Crassolabium* Yeates, 1967 (Figs. 3.108 and 3.109)

Diagnosis: These are moderately slender nematodes of medium size. The lip region is angular, offset by a marked depression; each lateral lip has two thickenings in the outer portion. The amphid fovea is funnel-like. The odontostyle is moderately robust. The guiding ring is simple. The pharyngeal gland nuclei and outlets are obscure. The genital system is didelphic-amphidelphic. The spicules are dorylaimoid. The supplements are composed of an adanal pair and a contiguous ventromedian series. The tails are convex and conoid to rounded and similar in both sexes.

Type and only species

Crassolabium australe Yeates, 1967

Family Nordiidae Jairajpuri and Siddiqi, 1964

Diagnosis: These are small to large nematodes. The cuticle has fine transverse striations. The lip region is offset or continuous with the body contour. The lips are fused or separate; the labial papillae are often raised or setiform. The amphids are stirrup shaped and the aperture is slit-like. The feeding apparatus

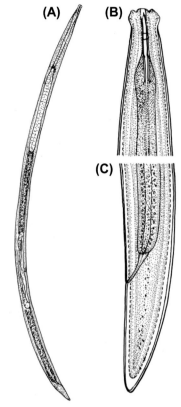

FIGURE 3.106 *Kochinema* sp.: (A) female; (B) anterior end; (C) posterior region.

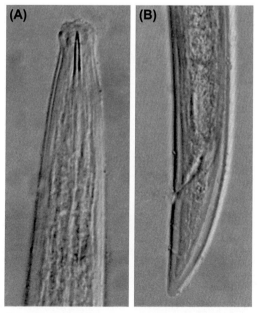

FIGURE 3.107 *Kochinema* sp.: (A) anterior end; (B) posterior region.

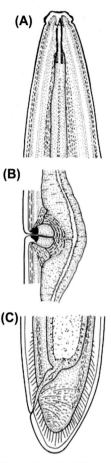

FIGURE 3.108 *Crassolabium* sp.: (A) anterior region; (B) vulval region; (C) female posterior region.

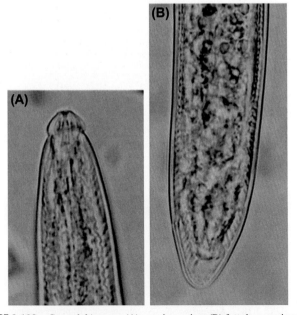

FIGURE 3.109 *Crassolabium* sp.: (A) anterior region; (B) female posterior region.

has a long, slender odontostyle with a very fine lumen and aperture, and a rod-like or basally flanged or swollen odontophore. The guiding ring is single or double. The anterior part of the pharynx is narrow; the basal part is expanded and muscular, and is usually less than half of the pharyngeal length. The females have a transverse vulva and a mono-opisthodelphic or amphidelphic genital system. The tail in both sexes is similar; the length is variable, hemispherical, and short and conoid to long and filiform.

Type genus

Longidorella Thorne, 1939

Key characteristics of some commonly found genera of Nordiidae

1. The cuticle is smooth; the inner labial papillae are setiform; females have an amphidelphic genital system . . . .*Cephalodorylaimus*
2. The cuticle is smooth; the inner labial papillae are barely raised; females have a monoprodelphic genital system . . . .*Acephalodorylaimus*
3. The labial papillae are not raised; the odontophore is usually flanged; four cuticularized pieces are present around the stoma. . . . . *Pungentus*
4. The labial papillae are not raised; the odontophore is usually flanged; four cuticularized pieces are absent around the stoma . . . . *Enchodelus*
5. The body is small; the amphids are narrow; the lip region is continuous; the odontostyle is usually more than twice the width of the lip region; the first pair of pharyngeal subventral glands are close together. . . .*Actinolaimoides*
6. The lip region is set off; the first pair of subventral pharyngeal glands are widely separated; the odontostyle length is less than two lip widths, the pharyngo-intestinal junction has well-developed glands. . . . . . .*Oriverutus*
7. The odontostyle length is three lip widths; the pharyngo-intestinal junction has no glands . . . .*Pararoriverutus*

Genera recorded

Pungentus Thorne and Swanger, 1936
Enchodelus Thorne, 1939

Genus *Pungentus* Thorne and Swanger, 1936
(Figs. 3.110 and 3.111)

Diagnosis: The body is small to medium-sized. The lips are angular with raised labial papillae. The lip region is offset. The stoma has four cuticularized pieces. The amphidial fovea is stirrup shaped and the aperture wide and slit-like. The odontostyle has a clear lumen and aperture. The odontophore is simple and rod-like. The basal expanded part of the pharynx occupies less than half of the pharyngeal length. The females have a transverse vulva and a mono-opisthodelphic or, rarely, an amphidelphic genital system. The spicules are arcuate and lateral

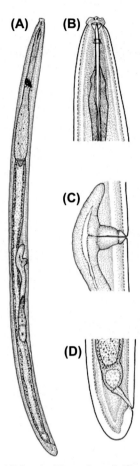

FIGURE 3.110 *Pungentus* sp.: (A) female; (B) anterior end; (C) vulval region; (D) posterior region.

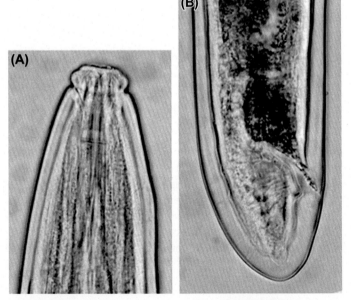

FIGURE 3.111 *Pungentus* sp.: (A) anterior end; (B) posterior region.

guiding pieces are present. The ventromedian supplements are spaced. The tails in both sexes are hemispherical or elongated and conoid, and sometimes clavate.

Type species

Pungentus pungens Thorne and Swanger, 1936

Commonly found species of *Pungentus* Thorne and Swanger, 1936

P. engadinensis (Altherr, 1950) Altherr, 1952
P. silvestris (De Man, 1912) Coomans and Geraert, 1962

Genus *Enchodelus* Thorne, 1939 (Figs. 3.112 and 3.113)

Diagnosis: The lip region is usually set off by a depression. The odontostyle is long and attenuated with a narrow lumen and aperture. The odontophore is

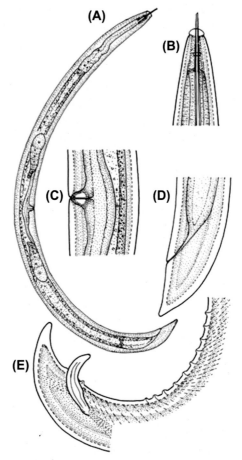

FIGURE 3.112 *Enchodelus* sp.: (A) female; (B) anterior end; (C) vulval region; (D) female posterior region; (E) male posterior region.

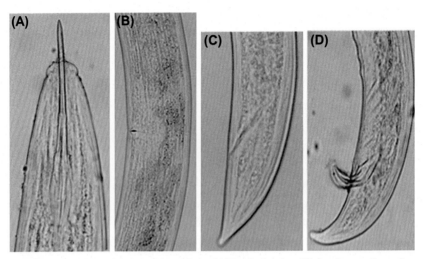

FIGURE 3.113 *Enchodelus* sp.: (A) anterior end; (B) vulval region; (C) female posterior region; (D) male posterior region.

simple and rodlike or it has basal flanges. The basal expanded part of the pharynx is relatively small. The females have an amphidelphic genital system. The prerectum is mostly long. The spicules are strong and arcuate; lateral guiding pieces are present. The number of ventromedian supplements varies from four to 12. The tail in both sexes is short or elongated and conoid to hemispheroid or with a rounded terminus.

Type species

Enchodelus macrodorus (De Man, 1880) Thorne, 1939

Commonly found species of *Enchodelus* Thorne, 1939

E. macrodorus (De Man, 1880) Thorne, 1939
E. constrictus Jairajpuri and Loof, 1968

Superfamily Longidoroidea Thorne, 1935

Diagnosis: These are usually long and slender nematodes; the body length is 1–12 mm. The lip region is set off from or continuous with the body. The fovea of amphids is a pouch or funnel, or it is stirrup shaped; the aperture is a pore or slit-like. The odontostyle is long and slender; the lumen is narrow and the aperture is very small. The odontophore may or may not have basal flanges. The junction of the odontostyle and odontophore is simple or complex. The anterior part of the pharynx is not muscular; usually it is convoluted when the odontostyle is retracted. The females have a monodelphic or amphidelphic genital system and a transverse vulva. The males have arcuate spicules, lateral guiding pieces, and ventromedian supplements. The tails in both sexes are similar.

Type family

Longidoridae Thorne, 1935*

Other family

Xiphinematidae Dalmasso, 1969*

Family Longidoridae Thorne, 1935

Diagnosis: The lip region is set off or continuous with the body. The fovea of the amphids is a pouch or funnel, or it is stirrup shaped; the aperture is a pore or slit-like. The guiding ring is single. The odontostyle–odontophore junction is simple. The odontophore is not flanged. The females have an amphidelphic genital system and a transverse vulva. The males have arcuate spicules, lateral guiding pieces, and ventromedian supplements. The tails in both sexes are short and conoid.

Type genus

Longidorus Micoletzky, 1922

Key characteristics of commonly found genera of Longidoridae

1. The amphids have pouch-like fovea and pore-like or inconspicuous apertures . . . . *Longidorus*
2. The amphids have pouch-like fovea and slit-like apertures . . . . *Longidoroides*
3. The amphids have stirrup-shaped fovea and large slit-like apertures . . . . *Paralongidorus*

Genus recorded

Longidorus Micoletzky, 1922

Genus *Longidorus* Micoletzky, 1922 (Figs. 3.114 and 3.115)

Diagnosis: These are long and slender nematodes. The lips are fused. The amphids have pouch-like fovea and small pore-like apertures. The odontostyle is long; the odontostyle–odontophore junction is simple. The guiding ring is single and located anteriorly. The odontophore is not flanged. The pharynx is typical of the family. The females have amphidelphic gonads. The males have arcuate spicules, lateral guiding pieces, and ventromedian supplements. The tails in both sexes are short and conoid.

Type species

Longidorus elongatus (De Man, 1876) Thorne and Swanger, 1936

Commonly found species of *Longidorus* Micoletzky, 1922

L. elongatus (De Man, 1876) Thorne and Swanger, 1936
L. africanus Merny, 1966

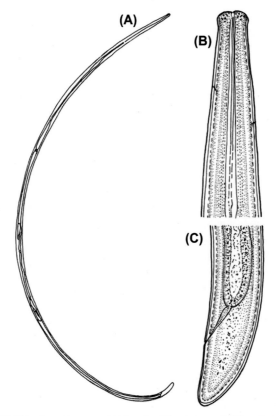

FIGURE 3.114 *Longidorus* sp.: (A) female; (B) anterior end; (C) posterior region.

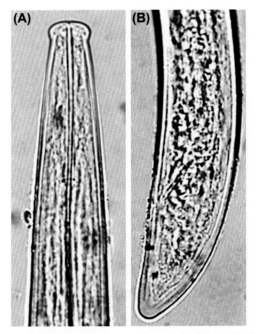

FIGURE 3.115 *Longidorus* sp.: (A) anterior end; (B) female posterior region.

Family Xiphinematidae Dalmasso, 1969

Diagnosis: The lip region is set off or is continuous with the body. The fovea of the amphids is a pouch or is stirrup shaped; the aperture is a pore or slit-like. A double guiding ring is present at the base of the odontostyle. The odontostyle is long and bifurcated at the junction with the odontophore. The odontophore base is moderately or well flanged. The females have a monodelphic or amphidelphic genital system. The males have arcuate spicules, lateral guiding pieces, and ventromedian supplements. The tails in both sexes are short and conoid to long and filiform.

Type genus

Xiphinema Cobb, 1913

Key characteristics of some common genera of Xipinematidae

1. The amphids have stirrup-shaped and slit-like apertures; the dorsal gland nucleus opens at the same level; subventral nuclei are less developed than the dorsal nucleus.................................*Xiphinema*
2. The amphids have pouch-shaped and pore-like apertures; the dorsal gland nucleus opens far behind; subventral nuclei are more developed than the dorsal nucleus.......................................*Xiphidorus*

Genus recorded

Xiphinema Cobb, 1913.

Genus *Xiphinema* Cobb, 1913 (Figs. 3.116 and 3.117)

Diagnosis: The nematodes are long and slender, up to 6 mm long. The lip region is offset or continuous with the body. The fovea of the amphid is stirrup shaped; the aperture is large and slit-like. A double guiding ring is present at the base of the odontostyle. The odontostyle is long and bifurcated at the junction with the odontophore. The base of the odontophore is well flanged. The anterior slender part of the pharynx is almost three-fourths the total esophageal length. There are only three pharyngeal glands. The dorsal esophageal gland nucleus (DN) and dorsal esophageal gland orifice are at the same level. The DN is more developed than the subventral esophageal gland nuclei. The females have monodelphic, pseudomonodelphic, amphidelphic genital systems. The males have arcuate spicules, lateral guiding pieces, and ventromedian supplements. The tails in both sexes are short and conoid, and are rarely long and filiform.

Type species

Xiphinema americanum Cobb, 1913

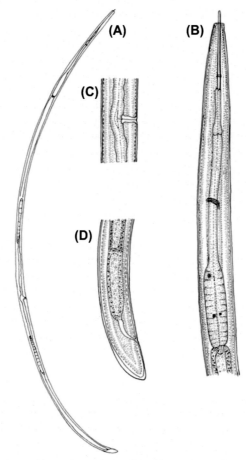

FIGURE 3.116 *Xiphinema* sp.: (A) female; (B) pharyngeal region; (C) vulval region; (D) female posterior region.

Commonly found species of *Xiphinema* Cobb, 1913

X. americanum Cobb, 1913
X. diversicaudatum (Micoletzky, 1927) Thorne, 1939
X. elongatum Schuurmans Stekhoven and Teunissen, 1938
X. index Thorne and Allen, 1950
X. insigne Loos, 1949

Superfamily Belondiroidea Thorne, 1939

Diagnosis: The lip region is generally narrow with small, angular, or rounded lips. The odontostyle is small; the odontophore is usually simple and rodlike;

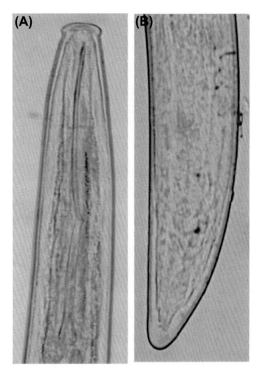

FIGURE 3.117   *Xiphinema* sp.: (A) anterior end; (B) female posterior region.

it is rarely flanged. The posterior expanded part of the pharynx is covered by spiral or, rarely, longitudinal muscle sheaths. The genital system in females is monodelphic or amphidelphic. The shape and size of spicules are variable. Ventromedian supplements are variable, from a few to many. The tail shape is variable and is similar or dissimilar in both sexes.

Type and only family

Belondiridae Thorne, 1939*

Family Belondiridae Thorne, 1939

Diagnosis: The lip region is generally narrow and offset or continuous. The odontostyle is small. The odontophore is usually simple and rodlike or flanged. The posterior expanded part of the pharynx is variable and continuous with the anterior slender part, or is set off by constriction. The shape of the cardia is variable. The genital system in females is monodelphic or amphidelphic. The shape and size of spicules are variable. Ventromedian supplements vary from a few to many. The tail shape is variable, and is similar or dissimilar in both sexes.

Type genus

Belondira Thorne, 1939

Key characteristics of some common genera of Belondiridae

1. Cuticularized pieces are present around the oral aperture; the odontophore is flanged. . . . . *Dorylaimellus*
2. The labial region is slightly sclerotized; pharyngeal expansion is gradual; the female genital system is mono-opisthodelphic; the female has a short, conoid, or rounded tail . . . . *Belondira*
3. The cephalic region has six liplets; pharyngeal expansion is gradual; the female genital system is amphidelphic . . . . *Belondirella*
4. The labial region is slightly sclerotized; the cephalic region has no liplets; pharyngeal expansion is gradual; the female genital system is amphidelphic; ventromedian supplements are absent. . . . .
. . . . *Amphibelondira*
5. The labial framework has no sclerotization; the odontostyle is fusiform to cylindroid; pharyngeal expansion is abrupt; the expanded part is very long
. . . . *Axonchium*
6. The odontostyle is small; the lip region is slightly asymmetrical; the tail is long and filiform in both sexes. . . . . *Oxydirus*

Genera recorded

Axonchium Cobb, 1920
Dorylaimellus Cobb, 1913

Genus *Axonchium* Cobb, 1920 (Figs. 3.118 and 3.119)

Diagnosis: These are small to large nematodes with a body length of 0.9–4.3 mm. They are almost straight or slightly curved ventrally upon relaxation. The amphids have slit-like apertures. The lip region is narrow and continuous or offset from the body by a constriction. The lips are conoid or rounded. The odontostyle is small and fusiform; the odontophore has thick walls. The anterior part of the pharynx is differentiated by an isthmus-like portion; the musculature is generally weak. The posterior expanded part of the pharynx is long and covered by spiral or longitudinal muscles. The females have a transverse, longitudinal, or, rarely, oval vulva and a mono-opisthodelphic genital system. The prevulval uterine sac is moderately long or short, or sometimes is completely absent. The males have large, proximally broad, and straight to arcuate spicules and bifid lateral guiding pieces. There are two to 30 ventromedian supplements that are contiguous or spaced. The tail in both sexes is hemispherical to conoid.

Type species

Axonchium amplicolle Cobb, 1920

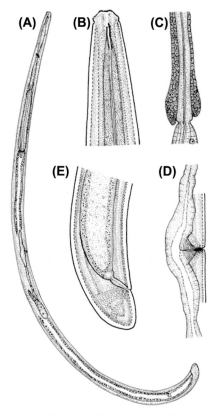

FIGURE 3.118 *Axonchium* sp.: (A) female; (B) anterior end; (C) junction of narrow and expanded part of pharynx; (D) vulval region; (E) female posterior region.

Commonly found species of *Axonchium* Cobb, 1920

A. amplicolle Cobb, 1920
A. coronatum (De Man, 1906) Thorne and Swanger, 1936
A. vulvulatum Nair and Coomans, 1974

Genus *Dorylaimellus* Cobb, 1913 (Figs. 3.120 and 3.121)

Diagnosis: These nematodes are small. The body is 0.4–1.8 mm in length and is usually curved ventrally upon relaxation. The cuticle has fine striations; glandular organs are present in the lateral chords. The amphids have slit-like apertures. The lip region is narrow and offset from the body by a constriction. The labial disc around the oral aperture may be present or absent; the labial papillae are raised above the labial contour. The oral aperture is surrounded by four cuticularized pieces. The odontostyle is small; the odontophore has basal flanges. The posterior expanded part of the pharynx is covered by a sheath of spiral muscles.

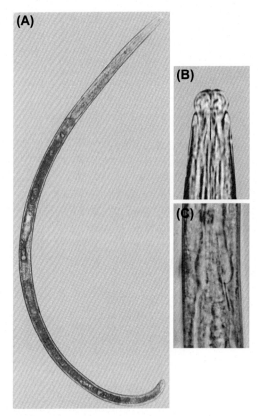

FIGURE 3.119   *Axonchium* sp.: (A) female; (B) anterior end; (C) junction of narrow and expanded part of pharynx.

The pharyngo-intestinal junction has a conoid or hemispheroid cardia. The females have a longitudinal but rarely transverse vulva and a monodelphic or amphidelphic genital system. The males have large, proximally broad, arcuate spicules; lateral guiding pieces; and usually spaced ventromedian supplements. The tail shape is variable and is similar in both sexes.

Type species

Dorylaimellus virginianus Cobb, 1913

Commonly found species of Dorylaimellus Cobb, 1913

D. tenuidens Thorne, 1957
D. virginianus Cobb, 1913

Superfamily Tylencholaimoidea Filipjev, 1934

Diagnosis: The cuticle has a loose subcuticle with fine or coarse striations, radial striations, and fixation folds. The lip region is set off from or continuous with the

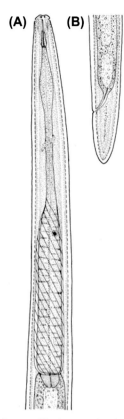

FIGURE 3.120 *Dorylaimellus* sp.: (A) pharyngeal region; (B) female posterior region.

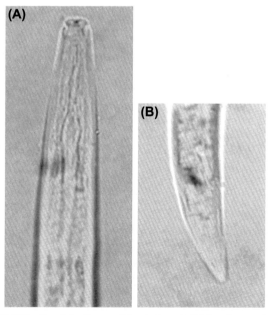

FIGURE 3.121 *Dorylaimellus* sp.: (A) anterior end; (B) female posterior region.

body, with or without a labial disc or inner liplets around the oral opening. The lips are mostly rounded; the papillae are not raised. The odontostyle is either symmetrical or asymmetrical, and is often solid and needle-like. The odonto-phore is simple and rodlike or slightly arcuate. Basal flanges or basal knobs are either present or absent. The pharynx may be slender with a small cylindrical or pyriform basal bulb, or it may be large and dorylaimoid. The posterior part of the pharynx may have a thickened lumen or it may have a well-developed triquetrous chamber. The females have a transverse or longitudinal vulva and a monodelphic or amphidelphic genital system. The spicules are arcuate; lateral guiding pieces are present; and ventromedian supplements are spaced and few in number. The tail has a variety of shapes and sizes, and is similar in both sexes.

Type family

Tylencholaimidae Filipjev, 1934*

Other families

Leptonchidae Thorne, 1935*
Aulolaimoididae Jairajpuri, 1964
Mydonomidae Thorne, 1964*

Family Tylencholaimidae Filipjev, 1934

Diagnosis: The cuticle is loose, with clear radial striations and fixation folds. The lip region is cap-like and is usually set off by a constriction. The lips are angular; labial papillae are not prominent. The labial disc around the oral open-ing is either present or absent. The odontostyle is usually small; it is sym-metrical or asymmetrical, with a clear lumen and aperture. The odontophore is rodlike, usually with basal knobs or flanges. The pharynx is dorylaimoid; the lumen of the basal expanded part of the pharynx is thickened. The females have a transverse or longitudinal vulva and a monodelphic or amphidelphic genital system. The spicules are well-developed and arcuate; lateral guiding pieces are present; and ventromedian supplements spaced and few in number. The tails in both sexes range from almost rounded to long and filiform.

Type genus

Tylencholaimus De Man, 1876

Key characteristics of some commonly found genera of Tylencholaimidae

1. The amphids have slit-like apertures; the odontophore has clear basal knobs; the tail is short to elongated and conoid. . . . . *Tylencholaimus*
2. A labial disc is present; the odontostyle is long with a narrow lumen and aperture; the odontophore is flanged; the females have an amphidelphic genital system; the tail is short and conoid . . . . *Xiphinemella*
3. The stoma has no sclerotization; the odontostyle is long with a narrow lumen and aperture; the odontophore is simple and rodlike, without knobs or flanges . . . . *Kantbhala*

4. The odontostyle walls are very thick and ventrally curved, and the base is furcated; the pharynx has two distinct parts; the vulva is transverse . . . . *Curvidorylaimus*

5. The odontostyle walls are very thick and ventrally curved, but the base is not furcated; the pharynx has indistinct parts; the vulva is circular . . . . *Vanderlindia*

6. The odontostyle is massive and has a small aperture; the vulva is transverse. . . . . *Metadorylaimus*

7. The amphids are small and pouch-like, and the aperture is oval; the odontostyle is asymmetrical; the female has a mono-opisthodelphic gonad; the tail is long and filiform . . . . *Mumtazium*

8. The amphids are small and pouch-like, and the aperture is oval; the odontostyle is asymmetrical; the female has amphidelphic gonads; the tail short and conoid. . . . . *Promumtazium*

Genus recorded

Tylencholaimus De Man, 1876.

Genus *Tylencholaimus* De Man, 1876 (Figs. 3.122 and 3.123)

Diagnosis: The body is generally less than 1.5 mm. Both the cuticle and subcuticle have fine transverse and radial striations. The lip region is set off from the body by deep constrictions. The odontostyle is almost one lip region long. The odontophore is rodlike and basal knobs are present. The pharynx is dorylaimoid with anterior slender and posterior expanded parts; pharyngeal expansion is either gradual or sudden. The females have a transverse vulva and monodelphic or amphidelphic gonads. The males have dorylaimoid spicules, lateral guiding pieces, and a few spaced ventromedian supplements. The tail in both sexes is hemispheroid to elongated and conoid.

Type species

Tylencholaimus mirabilis (Butschli, 1873) De Man, 1876

Commonly found species of *Tylencholaimus* De Man, 1876

T. proximus Thorne, 1939
T. teres Thorne, 1939

Family Leptonchidae Thorne, 1935

Diagnosis: Both the cuticle and subcuticle have fine transverse and radial striations, rarely with longitudinal lamelliform patterns. The lateral sector has two rows of body pores. The lip region is set off from the body, with or without a labial disc around the oral opening. The amphids are cyathiform, usually with a thick wall; the fovea is single or duplex. The odontostyle is slender, often

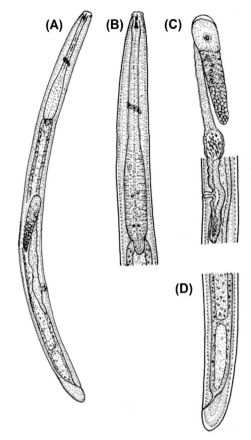

FIGURE 3.122 *Tylencholaimus* sp.: (A) female; (B) pharynx; (C) female genital system; (D) female posterior region.

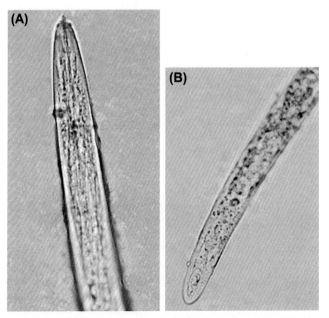

FIGURE 3.123 *Tylencholaimus* sp.: (A) pharynx; (B) female posterior region.

solid and pointed. The odontophore is simple and rodlike or arcuate, or it has basal flanges. The anterior part of the pharynx is slender; the posterior part is cylindroid or pyriform, or it is represented by a small basal bulb with or without a thickened inner cuticular lining. The pharyngo-intestinal junction has a hemispheroid or elongated cardia. The females have a transverse or longitudinal vulva and a monodelphic or amphidelphic genital system. The males have well-developed, arcuate spicules and a few spaced ventromedian supplements. Lateral guiding pieces are present. The tails in both sexes are rounded or long and filiform.

Type genus

Leptonchus Cobb, 1920

Key characteristics of some commonly found genera of Leptonchidae

1. The labial disc is absent; inner liplets are present around the oral aperture; the odontostyle has a wide lumen and aperture; the inner lumen of the pharyngeal bulb is distinctly thickened . . . . *Tyleptus*
2. The inner liplets around the oral aperture are absent; the pharyngeal bulb has no distinct valvular chamber . . . . *Gymnotyleptus*
3. The stoma has a weakly sclerotized base; the odontostyle is solid and needle-like; the odontophore is rodlike or slightly flanged. . . . . *Basirotyleptus*
4. The lip region is set off; the odontostyle is slender; the odontophore is arcuate and sclerotized; pharyngeal expansion is gradual; female gonads are amphidelphic . . . . *Leptonchus*
5. The lip region is slightly set off; the odontophore is flanged; the female genital system is monoprodelphic; the tail is digitate-spicate . . . . *Proleptonchoides*
6. The odontophore is not flanged; pharyngeal expansion is sudden; the female gonad is monoprodelphic; the tail is hemispherical or bluntly conoid . . . . *Proleptonchus*
7. The odontophore is straight and sclerotized only at the base, the vulva is pore-like . . . . *Apoleptonchus*

Genera recorded

Tyleptus Thorne, 1939
Leptonchus Cobb, 1920

Genus *Tyleptus* Thorne, 1939 (Figs. 3.124 and 3.125)

Diagnosis: The body is cylindroid and arcuate upon relaxation. The cuticle and subcuticle are smooth or have fine or coarse striae and many refractive radial elements. The lateral cords have two lines of lateral pores. The lip region is set off from the body; the inner liplets around the oral opening are conspicuous. The amphids have cup- or stirrup-shaped single or duplex

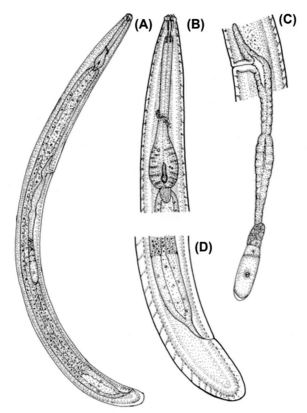

FIGURE 3.124   *Tyleptus* sp.: (A) female; (B) pharyngeal region; (C) female genital system; (D) female posterior region.

fovea. The odontostyle is strong and the odontophore is rodlike or flanged. The pharynx is slender, ending in a pyriform terminal bulb. The lumen of the terminal bulb is divisible into two parts: an anterior thickened portion and the posterior, forming a triquetrous, valvular chamber. The females have a transverse vulva and a mono-opisthodelphic genital system. The males have well-developed, dorylaimoid spicules, lateral guiding pieces, and a few spaced ventromedian supplements. The tails in both sexes are rounded, hemispheroid, or conoid.

Type species

Tyleptus projectus Thorne, 1939

Commonly found species of *Tyleptus* Thorne, 1939

T. projectus Thorne, 1939
T. striatus Heyns, 1963

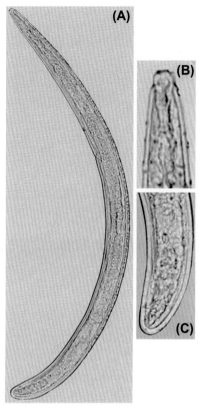

FIGURE 3.125 *Tyleptus* sp.: (A) female; (B) anterior region; (C) female posterior region.

Genus *Leptonchus* Cobb, 1920 (Figs. 3.126 and 3.127)

Diagnosis: These are usually small to medium-sized nematodes 0.7–1.5 mm in length. The body is arcuate upon fixation. The cuticle has fine striations and the subcuticle has coarse ones. The body pores are arranged in two rows near the lateral fields. The lip region is set off from the body and is usually cap-like. The odontostyle is cylindrical with a narrow. straight lumen. The odontophore is arcuate. The pharynx is slender, terminating in a pyriform bulb. The females have a transverse vulva and an amphidelphic genital system. The prerectum is usually long, sometimes extending beyond the vulva. The junction between the prerectum and intestine has three cells. The males have dorylaimoid spicules, lateral guiding pieces, and a few spaced ventromedian supplements. The tail in both sexes is hemispheroid to blunt conoid.

Type species

Leptonchus granulosus Cobb, 1920

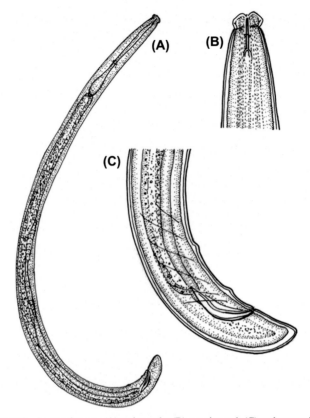

FIGURE 3.126 *Leptonchus* sp.: (A) entire male; (B) anterior end; (C) male posterior region.

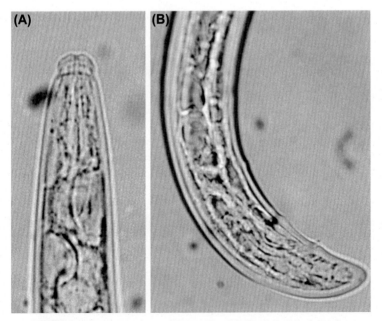

FIGURE 3.127 *Leptonchus* sp.: (A) anterior end; (B) male posterior region.

Commonly found species of *Leptonchus* Cobb, 1920

L. granulosus Cobb, 1920

Family Mydonomidae Thorne, 1964

Diagnosis: The cuticle and subcuticle have fine striations. The lips are rounded and the lip region is continuous with the body contour or is set off from the body. The odontostyle is asymmetrical; the odontophore is usually arcuate but sometimes straight. The basal expanded part of the esophagus is elongated or cylindroid. The females have a mono-opisthodelphic or amphidelphic reproductive system. The males have strong, arcuate spicules and lateral guiding pieces. Ventromedian supplements are few and spaced. The tail is short and conoid to long and filiform, and may or may not be similar in both sexes.

Type genus

Mydonomus Thorne, 1964

Key characteristics of some commonly found genera of Mydonomidae

1. The basal expanded part of the esophagus is enclosed in a muscular sheath
 . . . .*Mydonomus*
2. The basal expanded part of the esophagus is not enclosed in a muscular sheath; the tail is similar in both sexes. . . . .*Dorylaimoides*
3. The basal expanded part of the esophagus is not enclosed in a muscular sheath; the tail is not similar in both sexes. . . . .*Morasia*

Genus recorded

Dorylaimoides Thorne and Swanger, 1936.

Genus *Dorylaimoides* Thorne and Swanger, 1936
(Figs. 3.128 and 3.129)

Diagnosis: These are medium-sized nematodes with a body less than 2 mm in length. Both the cuticle and subcuticle have fine striations. The lip region is slightly set off from the body. The odontostyle is not symmetrical; the odontophore is arcuate and sometimes angular, usually enclosed in pharyngeal tissue. The basal expanded part of the pharynx is cylindroid, one-fourth to one-third of the total pharyngeal length. The females have a mono-opisthodelphic or amphidelphic reproductive system and a transverse vulva. The males have arcuate spicules and lateral guiding pieces. Ventromedian supplements are spaced. The tail in both sexes is hemispheroid to long and filiform.

Type species

Dorylaimoides teres Thorne and Swanger, 1936

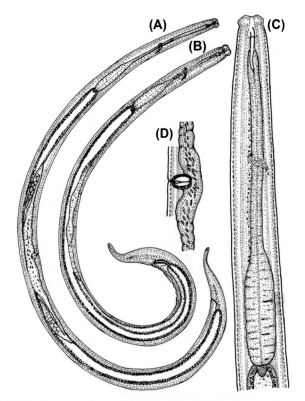

FIGURE 3.128 *Dorylaimoides* sp.: (A) female; (B) male; (C) pharynx; (D) vulval region.

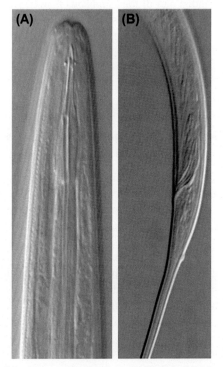

FIGURE 3.129 *Dorylaimoides* sp.: (A) anterior end; (B) male posterior region. *Photographed by Wasim Ahmad.*

Commonly found species of *Dorylaimoides* Thorne and Swanger, 1936

D. teres Thorne and Swanger, 1936
D. buccinator Sauer, 1967
D. chathami Yeates, 1979.

Suborder Nygolaimina Ahmad and Jairajpuri, 1979

Diagnosis: The stoma has a mural tooth on the left subventral wall. The stoma is divisible into distal, median, and proximal parts. The distal part has a thin wall; the median and proximal parts have thick walls. The pharyngo-intestinal junction has a cardiac disc or three cardiac glands. The females have a transverse or, rarely, longitudinal vulva and an amphidelphic or, rarely, mono-opisthodelphic genital system. The males have well-developed spicules, lateral guiding pieces, and ventromedian supplements. The gubernaculum may be either present or absent. The tail shape and size are variable but are similar in both sexes.

Type and only superfamily

Nygolaimoidea Thorne, 1935

Superfamily Nygolaimoidea Thorne, 1935

Diagnosis: The stoma has a mural tooth on the left subventral wall. The pharynx has anterior slender and posterior expanded parts. The posterior part of the pharynx is sometimes bibulbular and generally is enclosed in a sheath. The pharyngo-intestinal junction has a cardiac disc or three cardiac glands. The females have a transverse or, rarely, longitudinal vulva and an amphidelphic or, rarely, mono-opisthodelphic genital system. The males have well-developed spicules, lateral guiding pieces, and ventromedian supplements. The gubernaculum may be either present or absent. The tail shape and size are variable and are similar in both sexes.

Type family

Nygolaimidae Thorne, 1935*

Other families

Nygellidae Andrássy, 1958
Nygolaimellidae Clark, 1961
Aetholaimidae Jairajpuri, 1965

Family Nygolaimidae Thorne, 1935

Diagnosis: The cuticle is usually thin; fine striations are present on the inner or outer layers, or both. The lip region is set off or continuous with both contours; the lips are largely fused. The shape and size of the mural tooth are variable. The basal expanded part of the pharynx is enclosed in a thin muscular sheath. The pharyngo-intestinal junction has three cardiac glands. The females have a transverse or, rarely, longitudinal vulva and an amphidelphic genital system. The males have well-developed spicules, lateral guiding pieces, and ventrome- dian supplements. The gubernaculum may be either present or absent. The tail shape and size are variable and are similar in both sexes.

Type genus

Nygolaimus Cobb, 1913

Key characteristics of some commonly found genera of Nygolaimidae

1. The mural tooth is deltoid or linear; the lip region is set off; the gubernaculum is absent; ventromedian supplements are weakly developed .... *Nygolaimus*
2. The mural tooth is deltoid or linear; the lip region is set off; the gubernaculum is present; ventromedian supplements are well developed. .*Paranygolaimus*
3. The tooth is dorylaimoid; the tail is short and convex and conoid, hemispheroid, or clavate............................ *Laevides*
4. The body is C or S shaped; the lip region is set off from the body by constriction .... *Solididens*
5. The body is more or less straight; the lip region is continuous with the body; the tooth is linear; the tail is convex and conoid to hemispheroidal .... *Aquatides*
6. The body is ventrally curved; the lip region is usually continuous with the body; the tooth is deltoid to linear; the tail is hemispheroidal to clavate .... *Clavicaudoides*

Genus *Nygolaimus* Cobb, 1913 (Figs. 3.130 and 3.131)

Diagnosis: The body length is variable from 1 to 4 mm. The cuticle is usu- ally thin; transverse striations are present on the inner and outer layers. The lip region is offset. The mural tooth is deltoid; the length is usually less than one lip region wide. The basal expanded part of the pharynx is enclosed in a thin mus- cular sheath. The pharyngo-intestinal junction has three well-developed cardiac glands and a small cardia. The females have a transverse vulva and an amphidel- phic genital system. The males have well-developed spicules and small lateral guiding pieces. Ventromedian supplements are poorly developed or absent. The gubernaculum is absent. The tail in both sexes is conoid or conical.

Type species

Nygolaimus pachydermatus Cobb, 1913

Commonly found species of *Nygolaimus* Cobb, 1913

N. brachyuris (De Man, 1880) Thorne, 1930

Genus *Clavicaudoides* Heyns, 1968 (Figs. 3.132 and 3.133)

Diagnosis: The body is generally less than 2 mm in length. The cuticle and subcuticle have fine transverse striations. The lip region is set off from or

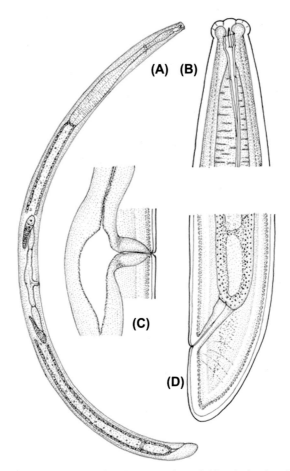

FIGURE 3.130 *Nygolaimus* sp.: (A) female; (B) anterior end; (C) vulval region; (D) female posterior region.

continuous with the body contour. The basal expanded part of the pharynx is enclosed in a thin muscular sheath. The pharyngo-intestinal junction has three well-developed ovoid or rounded cardiac glands. The females have a transverse vulva and an amphidelphic genital system. The males have small spicules and a few ventromedian supplements. The gubernaculum is absent. The tail in both sexes is clavate to hemispheroid.

Type species

Clavicaudoides clavicaudatus (Altherr, 1953) Ahmad and Jairajpuri, 1982

Commonly found species of *Nygolaimus* Cobb, 1913

C. altherri Heyns, 1968
C. clavicaudatus (Altherr, 1953) Ahmad and Jairajpuri, 1982

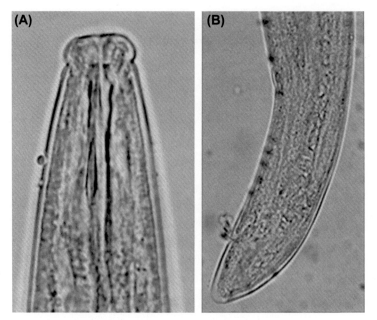

FIGURE 3.131 *Nygolaimus* sp.: (A) anterior ends; (B) female posterior region.

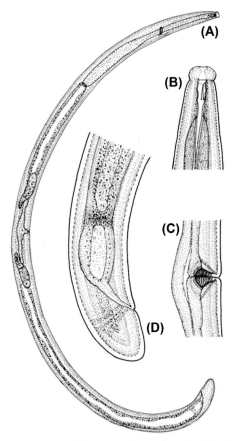

FIGURE 3.132 *Clavicaudoides* sp.: (A) female; (B) anterior end; (C) vulval region; (D) female posterior region.

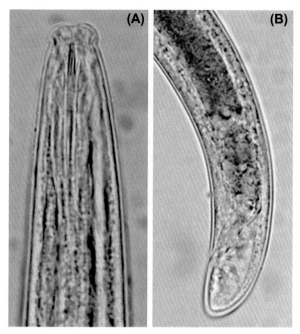

FIGURE 3.133   *Clavicaudoides* sp.: (A) anterior end; (B) female posterior region.

Suborder Campydorina Jairajpuri, 1983

Diagnosis: The feeding apparatus is provided with a hollow, acute mural tooth placed subdorsally. The pharynx muscular has an oblong basal bulb housing a large triquetrous chamber. An excretory pore and excretory duct present. The female genital system is amphidelphic with reflexed ovaries usually extending beyond the vulva.

Type and only superfamily

Campydoroidea Thorne, 1935

Superfamily Campydoroidea Thorne, 1935

Diagnosis: The cuticle is thin. The lip region is slightly set off from the body contour. The lips are separated with prominent labial papillae. The feeding apparatus is provided with a hollow, acute mural tooth placed subdorsally. The pharynx is muscular with an oblong basal bulb housing a large triquetrous chamber. An excretory pore and excretory duct are present. The female genital system is amphidelphic with reflexed ovaries usually extending beyond the vulva.

Type and only family

Campydoridae Thorne, 1935

Family Campydoridae Thorne, 1935

Diagnosis: The cuticle is thin. The lip region is slightly set off from the body contour. The lips are separated with prominent labial papillae. The feeding apparatus is provided with a hollow, acute mural tooth placed subdorsally. The pharynx is muscular with an oblong basal bulb housing a large triquetrous chamber. An excretory pore and excretory duct present. The female genital system is amphidelphic with reflexed ovaries usually extending beyond the vulva. The tail is elongated.

Type and only genus

Campydora Cobb, 1920

Genus *Campydora* Cobb, 1920 (Figs. 3.134 and 3.135)

Diagnosis: This is a small nematode, less than 1 mm. The cuticle is thin. The lip region is slightly set off from the body contour. The lips are separated

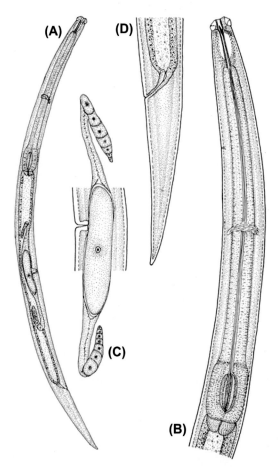

FIGURE 3.134 *Campydora* sp.: (A) female; (B) pharynx; (C) female genital system; (D) female posterior region.

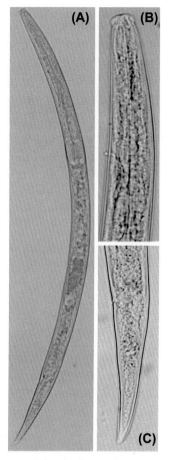

FIGURE 3.135 *Campydora* sp.: (A) female; (B) anterior pharynx; (C) female posterior region.

with prominent labial papillae. The feeding apparatus has a hollow, acute mural tooth placed subdorsally. The pharynx is muscular with a long, cylindrical muscular anterior part and an oblong, bulb-like basal part housing a large triquetrous chamber. An excretory pore and excretory duct are present. The female genital system is amphidelphic with reflexed ovaries usually extending beyond the vulva. The tail is elongated and the terminus is pointed.

Type and only species

Campydora demonstrans Cobb, 1920

Order Triplonchida Cobb, 1920

Diagnosis: These are usually short and stout nematodes with a body length of 1 mm or less. The cuticle is thick and loosely covers the body, with or without a few lateral pores and a pair of caudal pores. The amphids are just behind the lips, with large fovea and oval or ellipsoidal apertures. The stoma is a tube-like spear complex. The stomal wall is lined with complex cuticular plates. The pharynx has a slender anterior part and an oblong or pyriform posterior part. An excretory pore is usually present. The females generally have an amphidelphic or sometimes a monoprodelphic genital system. There is a single testis. Spicules are triploid. There is a gubernaculum. Ventromedian supplements are few and either weakly or well developed. The tail in both sexes is short.

Type suborder

Diphtherophorina Coomans and Loof, 1970

Other suborder

Tobrilina Tsalolikhin, 1976

Suborder Diphtherophorina Coomans and Loof, 1970

Diagnosis: Same as Triplonchida.

Type superfamily

Diphtherophoroidea Micoletzky, 1922*

Other superfamily

Trichodoroidea Thorne, 1935

Superfamily Diphtherophoroidea Micoletzky, 1922

Diagnosis: The body is short and plump, usually less than 1 mm in length. The stomal wall is lined with a series of complex cuticular plates. The spear has basal swelling. The pharynx is composed of a slender anterior part and an oblong or pyriform posterior part. An excretory pore is usually present. The females have an amphidelphic genital system. There is a single testis. Spicules are triploid and slightly ventrally arcuate. The gubernaculum is weakly developed. Ventromedian supplements are weakly developed. The tail in both sexes is short.

Type and only family

Diphtherophoridae Micoletzky, 1922

Diagnosis: Same as given previously.

Type genus

Diphtherophora De Man, 1880

Key characteristics of some common genera of Diphtherophoridae

1. The posterior expanded part of the pharynx has an elongated pyriform bulb; the dorsal side of the spear is short, divided and strongly curved . . .
. . . . *Diphtherophora*
2. The posterior expanded part of the pharynx has an elongated pyriform bulb; the dorsal side of the spear is pointed, straight, and not divided . . . .
. . . . *Tylolaimophorus*
3. The posterior expanded part of the pharynx is almost half of its length
. . . . *Longibulbophora*

Genus recorded

Diphtherophora De Man, 1880

Genus *Diphtherophora* De Man, 1880 (Figs. 3.136 and 3.137)

Diagnosis: This is a small nematode, less than 1 mm in length. The cuticle is loose and the lips are separated; the labial papillae are raised or represented by circular spots. The stomal wall is lined with a series of complex cuticular plates. The spear is strongly developed and complex; the basal swelling is knob-like. The anterior part of the pharynx is slender and the posterior part is oblong. An excretory pore is visible. The females have an amphidelphic genital system; the ovaries are reflexed. There is a single testis. The spicules are triploid and slightly ventrally arcuate. The gubernaculum is small. Ventromedian supplements are weakly developed. The tail in both sexes is short.

Type species

Diphtherophora communis De Man, 1880

Commonly found species of *Diphtherophora* De Man, 1880

D. communis De Man, 1880
D. obesa Thorne, 1939

Suborder Tobrilina Tsalolikhin, 1976

Diagnosis: The pharynx is cylindrical with three large precardiac glands present. The males have more than five supplements. The proximal ends of spicules are embedded in muscles (confirmed in some species of both *Prismatolaimus* and *Tobrilus*).

This classification is based on De Ley and Blaxter (2002), who found the superfamilies Prismatolaimoidea and Tobriloidea to form an independent clade in triplonchs. They based their claim on the findings of a study of small subunit ribosomal DNA.

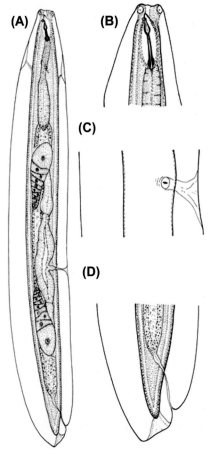

FIGURE 3.136   *Diphtherophora* sp.: (A) female; (B) anterior region; (C) vulval region; (D) posterior region.

Type superfamily

Prismatolaimoidea Micoletzky, 1922*

Other superfamily

Tripyloidea (Örley, 1880) Chitwood, 1937*

Superfamily Prismatolaimoidea Micoletzky, 1922

Diagnosis: These generally have wide, prismatic stoma; the tail is filiform; the tail terminus has a mucro and is similar in both sexes.

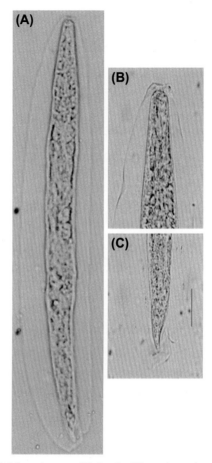

FIGURE 3.137   *Diphtherophora* sp.: (A) female; (B) anterior region; (C) posterior region.

Type family

Prismatolaimidae Micoletzky, 1922

Family Prismatolaimidae Micoletzky, 1922

Diagnosis: The cuticle has fine annulations and the annules are smooth. Lateral fields, body pores, and epidermal glands are absent. Somatic setae are present. The lip region is continuous with the body. The inner labial papillae are not modified into the setae; the outer labial papillae are setiform. The cephalic setae are well developed. The amphidial apertures are slit-like. The stoma is wide and cylindrical, and narrows posteriorly. The pharynx is muscular and tubular, gradually expanding posteriorly to join the pharyngo-intestinal junction. The dorsal pharyngeal gland opens near the base of the stoma. The pharyngo-intestinal junction has an ovoid, glandular cardia. The females have amphidelphic or monoprodelphic gonads. The ovary/ies are reflexed. The males have a pair of testes, spicules, and gubernaculum and a row of midventral supplements. The tail is elongated and narrows posteriorly; the tail tip has a mucro.

Type and only genus

Prismatolaimus De Man, 1880

Genus *Prismatolaimus* De Man, 1880 (Figs. 3.138 and 3.139)

Diagnosis: The cuticle has fine annulations and the annules are smooth. Lateral fields, body pores, and epidermal glands are absent. Somatic setae are present. The lip region is continuous with the body. The inner labial papillae are not modified into the setae; the outer labial papillae are setiform. The cephalic setae are well developed and slightly shorter than the outer labial setae. The

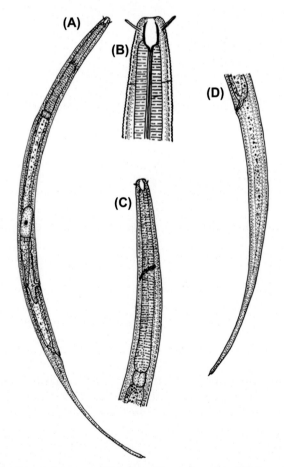

FIGURE 3.138 *Prismatolaimus* sp.: (A) female; (B) anterior region; (C) pharyngeal region; (D) posterior region.

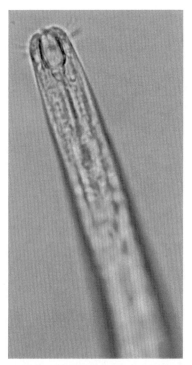

FIGURE 3.139   *Prismatolaimus* sp.: Anterior region.

amphidial apertures are slit-like. The stoma is wide and cylindrical, and narrows posteriorly. The pharynx is muscular and tubular, gradually expanding posteriorly to join the pharyngo-intestinal junction. The dorsal pharyngeal gland opens near the base of the stoma. The pharyngo-intestinal junction has an ovoid glandular cardia. The females have amphidelphic or monoprodelphic gonads. The ovary/ies are reflexed. The vulva is a transverse slit that is equatorial. The males have a pair of testes, spicules, a gubernaculum, and a row of midventral supplements. Midventral supplements are few to many (7–90), usually arranged in two groups. The tail is elongated, narrowing posteriorly; the tail tip has a mucro. Caudal glands and a spinneret are absent.

Type species

Prismatolaimus intermedius (Butschli, 1873) De Man, 1880

Commonly found species of *Prismatolaimus* De Man, 1880

P. aquaticus Daday, 1897
P. dadayi Stefanski, 1925
P. dolichurus De Man, 1988
P. intermedius (Butschli, 1873) De Man, 1880
P. verrucosus Hirschmann, 1952
P. waipukea Yeates, 1967

Order Mononchida Jairajpuri, 1969

Diagnosis: These are large and stout nematodes with a thick cuticle. The lip region is expanded with well-developed lips and labial papillae. The walls of the buccal cavity are strongly sclerotized; the posterior one-fourth or the whole buccal cavity is surrounded by pharyngeal tissue. The buccal cavity has a tooth or teeth; longitudinal ventral ridges are either present or absent; teeth or denticles are either present or absent on the subventral walls. The amphids are usually goblet shaped; the apertures are postlabial. The pharynx is a long, strongly muscular cylinder with strongly thickened lumen. There are five pharyngeal glands: uninucleate, dorsal and two subventral pairs. The females have a mono-prodelphic or amphidelphic genital system. The ovaries are reflexed. The males have identical spicules, a gubernaculum, accessory pieces, and ventromedian supplements. An adcloacal pair of supplements is absent. There are three caudal glands with or without a sclerotized terminal or subterminal spinneret.

Type suborder

Mononchina Kirjanova and Krall, 1969*

Other suborder

Bathyodontina Coomans and Loof, 1970

Suborder Mononchina Kirjanova and Krall, 1969

Diagnosis: These large and stout nematodes have a thick cuticle. The buccal cavity is spacious. The walls of the buccal cavity are strongly sclerotized; the posterior one-fourth is surrounded by pharyngeal tissue. The buccal cavity has a prominent dorsal tooth. Denticles on the ventral ridges are either present or absent. The subventral wall has teeth as large as the dorsal tooth, or it may have denticles, or both. The pharynx is a long, strongly muscular cylinder with a strongly thickened lumen. The spicules are strong, the gubernaculum is well developed, and accessory pieces are prominent.

Type superfamily

Mononchoidea Chitwood, 1937*

Other superfamily

Anatonchoidea Jairajpuri, 1969

Superfamily Mononchoidea Chitwood, 1937

Diagnosis: The buccal cavity has thick walls narrowing at the base. The dorsal tooth is medium- to large-sized. When subventral teeth are present, they are smaller than or as large as the dorsal tooth. When it is present, the ventral ridge may or may not have denticles. The pharyngo-intestinal junction is nontuberculate.

Type family

Mononchidae Chitwood, 1937

Other families

Mylonchulidae Jairajpuri, 1969*
Cobbonchidae Jairajpuri, 1969

Family Mylonchulidae Jairajpuri, 1969

Diagnosis: The dorsal tooth is in the anterior half or midway into the buccal cavity. One row or several rows of denticles are arranged transversely or scattered on the opposite wall of the dorsal tooth; subventral teeth may be present. The tail is short and conoid with caudal glands and a well-developed spinneret. Sometimes the spinneret is absent.

Type genus

Mylonchulus (Cobb, 1916) Altherr, 1953

Genus recorded

Nine genera are recognized under the family Mylonchulidae Jairajpuri, 1969, viz., *Brachonchulus* Andrássy, 1958; *Polyonchulus* Mulvey and Jensen, 1967; *Margaronchulus* Andrássy, 1972; *Oligonchulus* Andrássy, 1976; *Megaonchulus* Jairajpuri and Khan, 1982; *Paramylonchulus* Jairajpuri and Khan, 1982; *Crestonchulus* Siddiqi and Jairajpuri, 2002; *Margaronchuloides* Ahmad and Jairajpuri, 2010, and *Mylonchulus* (Cobb, 1916) Altherr, 1953. *Mylonchulus* is the only genus that was present in our collection and it is the only genus from this family that is widely distributed throughout the world.

Genus *Mylonchulus* (Cobb, 1916) Altherr, 1953 (Figs. 3.140 and 3.141)

Diagnosis: The buccal cavity narrows at the base and is usually goblet shaped. The dorsal tooth is large; it is present in the anterior half of the buccal cavity, usually opposed by a pair of ventral teeth at the base. Subventral walls have four or more transverse rows of denticles. The pharyngo-intestinal junction is nontuberculate. The females have an amphidelphic genital system; the ovaries are reflexed. The spicules are usually short. The gubernaculum is simple; the lateral accessory process is either present or absent. The tail shape varies.

Type species

Mylonchulus minor (Cobb, 1893) Andrássy, 1958

Commonly found species of *Mylonchulus* (Cobb, 1916) Altherr, 1953

M. brachyuris (Bulschli, 1873) Cobb, 1917
M. sigmatures Cobb, 1917

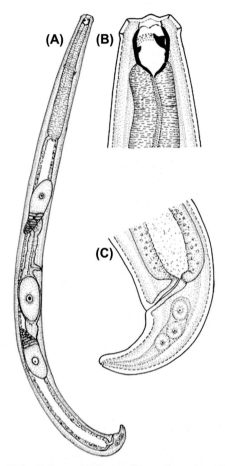

FIGURE 3.140 *Mylonchulus* sp.: (A) female; (B) anterior region; (C) posterior region.

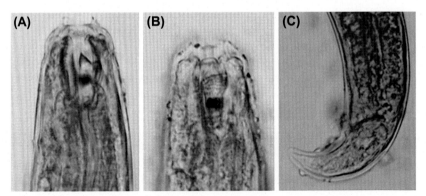

FIGURE 3.141 *Mylonchulus* sp.: (A) anterior region showing dorsal tooth; (B) anterior region showing rows of denticles; (C) posterior region.

Order Alaimida Siddiqi, 1983

Diagnosis: The cuticle is thin and smooth or it has fine striations. The somatic setae and metanemes may be absent (suborder Alaimina) or present (suborder Oxystominina). The lips have papillae or setae (suborder Oxystominina). The cephalic sensillae are variable: papillose or cupped. The amphids are located far posteriorly; the amphidial apertures are pore-like, circular, crescent shaped, transverse slit-like, elliptical, or extraordinarily elongated. The lip region is usually continuous with the body, and the lips are fused. The stoma is not well developed. The stomal walls have weak sclerotization. The pharynx is muscular and divisible into anterior narrow and posterior broad parts. The pharyngeal glands open in either a basal expanded part or an anterior nerve ring. Cardiac glands at the pharyngo-intestinal junction are absent. The females usually have a mono-opisthodelphic gonad; rarely, they have monoprodelphic or amphidelphic gonads. The prerecum is absent. Caudal glands and a spinneret may be present (Oxystominina) or absent (Alaimina). The males are monorchic (Alaimina) or diorchic (Oxystominina). The gubernaculum is either present (Oxystominina) or absent (Alaimina). The tail in both sexes is short to elongated and conoid to filiform.

Type suborder

Alaimina Clark, 1961*

Other suborder

Oxystominina Siddiqi, 1983

Suborder Alaimina Clark, 1961

Diagnosis: The lips are provided with papillae or are represented by small pit-like depressions. The cephalic sensillae are papillose. The amphids are located far posteriorly; the amphidial apertures are pore-like, circular, transverse slit-like, or crescent-shaped. The lip region is usually continuous with the body; the lips are fused. The stoma is not well developed or is vestigial, with no armature. The pharynx is muscular, expanding posteriorly, more prominently behind the nerve ring. The pharyngeal glands open into the basal expanded part. The females have a mono-opisthodelphic gonad; rarely, they have monoprodelphic or amphidelphic gonads. The prerectum is absent. The males have a single outstretched testis, ventromedian precloacal supplements, and straight or arcuate spicules. The gubernaculum is absent. The tail in both sexes is short to elongated and conoid to filiform.

Type and only superfamily

Alaimoidea Micoletzky, 1922

Superfamily Alaimoidea Micoletzky, 1922

Diagnosis: The amphids have pore- or crescent-like apertures located far posteriorly. The stoma is very small or vestigial, with no armature. The pharynx is muscular, expanding more prominently in the posterior one-third or less. The pharyngeal glands open into the basal expanded part. The females have a mono-opisthodelphic gonad, or rarely, monoprodelphic or amphidelphic gonads. The males have a single outstretched testis, ventromedian precloacal supplements, and straight or arcuate spicules. Adcloacal supplements are absent.

Type and only family

Alaimidae Micoletzky, 1922

Family Alaimidae Micoletzky, 1922

Diagnosis: The cuticle is smooth or, rarely, has longitudinal ridges. The amphids have pore-like or slit-like apertures. The lip region is continuous with the body. The stoma is vestigial with no armature. The pharynx is muscular; posterior expansion is gradual or it has sudden expansion in the posterior fifth to seventh. The female genital system is mono-opisthodelphic or amphidelphic. Ventromedian supplements are few to many. The length of spicules varies from very short to long.

Type genus

Alaimus De Man, 1880

Genus recorded

Only the type genus was found in our collection.

Key characteristics of some commonly found genera of Alaimidae

1. The cuticle has no longitudinal ridges; the amphidial apertures are pore-like; the spicules are short and simple . . . . *Alaimus*
2. The cuticle has longitudinal ridges; the amphidial apertures are indistinct; the spicules are short, simple, and straight, and the proximal end is large
. . . . *Cosalaimus*
3. The cuticle has fine transverse striations and several longitudinal ridges; the amphidial apertures are transverse; the vagina is S shaped; the spicules are slender. . . . . *Cristamphidelus*
4. The amphidial apertures are large, near the cephalic region; the vagina is normally thick-walled, not infolded. *Amphidelus*
5. The vaginal wall is infolded. . . . . *Metamphidelus*

Genus *Alaimus* De Man, 1880 (Figs. 3.142 and 3.143)

Diagnosis: The cuticle is smooth. The amphids have pore-like apertures. The lip region is continuous with the body or is more or less conical. The lips have minute labial papillae. The stoma is vestigial or small, with no armature. The pharynx is muscular, gradually expanding posteriorly. There are five to seven pharyngeal glands. The females have a mono-opisthodelphic genital system and a transverse vulva. A prevulval uterine sac is rarely present. The males have short and simple spicules and three to nine ventromedian supplements (rarely 16).

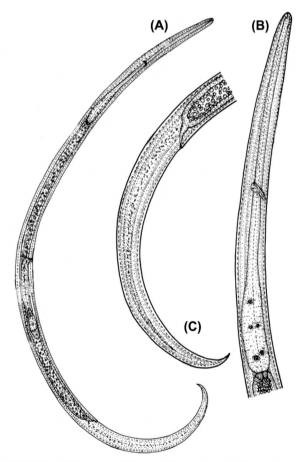

FIGURE 3.142 *Alaimus* sp.: (A) female; (B) anterior end; (C) posterior region.

FIGURE 3.143 *Alaimus* sp.: Anterior end.

REFERENCES

Ahmad, M., Jairajpuri, M.S., 1979. On the systematic position of Nygolaimina new suborder (Nematoda: Dorylaimida). In: Second National Congress of Parasitology. B.H.U, India, p. 29 (abstract).

Ahmad, M., Jairajpuri, M.S., 1982. Nygolaimina of India. Records of Zoological Survey of India, Occasional Paper, No. 34. 70 pp.

Andrássy, I., 1954. Revision der gattung *Tylenchus* Bastian, 1865 (Tylenchidae, Nematoda). Acta Zoologica Academiae Scientiarum Hungaricae 1, 5–42.

Andrássy, I., 1958. Szabadonélö fonálférgek (Nematoda libera). Fauna Hungariae 36, 392.

Andrássy, I., 1959. Taxonomische Übersicht der Dorylaimen (Nematoda). Journal of Acta Zoologica Academiae Scientiarum Hungaricae 5, 191–240.

Andrássy, I., 1968. Fauna Paraguayensis: 2. Nematoden aus den Galeriewäldern des Acaray-Flusses. Opuscula Zoologica Budapest 8, 167–315.

Andrássy, I., 1974. A Nematodák evolúciója és rendszerezése. Magyar Tudomanyos Akademia Biológiai Tudomanyok Osztályának Közleményei 17, 13–58.

Andrássy, I., 1981. Revision of the order Monhysterida (Nematoda) inhabiting soil and inland waters. Opuscula Zoologica Budapest 17/18, 13–47.

Andrássy, I., 1986. The genus *Eudorylaimus* Andrássy, 1959 and the present status of its species (Nematoda: Qudsianematidae). Opuscula Zoologica Budapestinensis 22, 1–42.

Andrássy, I., Zombori, L., 1976. Evolution as a basis for the systematization of nematodes. Journal of Parasitology 288 pp.

Barrett, J., 1991. Anhydrobiotic nematodes. Agricultural Zoology Review 4, 161–176.

Bastian, H.V., 1865. Monograph on the Anguillulidae, or free nematoids, marine, land and freshwater: with descriptions of 100 new species. Transactions of the Linnean Society of London 25, 73–184.

Bongers, T., 1994. De Nematoden van Nederland: Vormgeving en technische realisatie. Uitgeverij Pirola, Schoorl, Netherlands.

Caveness, F.E., 1964. A Glossary of Nematological Terms. Ibadan, Nigeria.

Chitwood, B.G., 1933. On some nematodes of the superfamily Rhabditoidea and their status as parasites of reptiles and amphibians. Journal of Washington Academy of Sciences 23, 508–520.

Chitwood, B.G., 1951. North American marine nematodes. Texas Journal of Science 3, 617–672.

Chitwood, B.G., 1958. The designation of official names for higher taxa of invertebrates. Bulletin of Zoological Nomenclature 15, 860–895.

Chitwood, B.G., Chitwood, M.B., 1950. General structure of nematodes. In: An Introduction to Nematology. Section 1, Anatomy. Monumental Printing Company, Baltimore, USA, pp. 7–27.

Clark, W.C., 1961. A revised classification of the order Enoplida (Nematoda). New Zealand Journal of Science 4, 123–150.

Cobb, N.A., 1893. Nematodes, mostly Australian and Fijian. In: McLeay Memorial Volume. Linnean Society of New South Wales, pp. 252–308.

Cobb, N.A., 1913. New nematode genera found inhabiting fresh water and non-brackish soils. Journal of the Washington Academy of Sciences 3, 432–444.

Cobb, N.A., 1914. Antarctic Marine Free-Living Nematodes of the Shakleton Expedition. Contributions to a Science of Nematology, vol. 1, pp. 1–33.

Cobb, N.A., 1916. Subdivisions of *Mononchus*. International Journal for Parasitology 2, 195–196.

Cobb, N.A., 1920. One Hundred New Nemas (Type Species of 100 New Genera). Contributions to a Science of Nematology, vol. 9, pp. 217–343.

Coomans, A., Loof, P.A.A., 1970. Morphology and taxonomy of Bathyodontina (Dorylaimida). Nematologica 16, 180–196.

Dalmasso, A., 1969. Etude anatomique at taxonomique des genres *Xiphinema*, *Londidorus* et *Paralongidorus* (Nematoda: Dorylaimidae). Mémoires Du Muséum National Dhistoire Naturelle Série A Zoologie 61, 33–82.

De Coninck, L.A., Schuurmans Stekhoven, J.H., 1933. The free living marine nemas of the Belgian coast. II, with general remarks on the structure and the system of nemas. Mémoires du Musée Royal d'Histoire Naturelle de Belgique 58, 1–163.

De Grisse, A., Loof, P.A.A., 1965. Revision of the genus *Criconemoides* (Nematoda). Journal Mededelingen van de Landbouwhogeschool en der Opzoekingsstations van de Staat te Gent 30, 577–603.

De Ley, P., Blaxter, M.L., 2002. Systematic position and phylogeny. In: Lee, D.L. (Ed.), The Biology of Nematodes. Taylor and Francis, London, pp. 1–30.

De Man, J.G., 1876. Onderzoekingen over vrij in de aarde levende Nematoden. Tijdschrift der Nederlandsche Dierkundige Vereeniging 2, 78–196.

De Man, J.G., 1880. Die einheimischen, frei in der Erde und im süssen wasser lebenden Nematoden. Vorläufiger Bericht und descriptive-systematischer Theil. Tijdschrift der Nederlandsche Dierkundige Vereeniging 5, 1–104.

De Man, J.G., 1921. Nouvelles recherches sur les nematodes libres terricoles de la Hollande. Capita Zoologica 1, 3–62.

Filipjev, I.N., 1934. The classification of the free-living nematodes and their relation to parasitic nematodes. Smithsonian Miscellaneous Collection (Washington) 89, 1–63.

Filipjev, I.N., 1936. On the classification of the Tylenchinae. Proceedings of the Helminthological Society 3, 80–82.

Freckman, D.W., Kaplan, D.T., Van Gundy, S.D., 1977. A comparison of techniques for extraction and study of anhydrobiotic nematodes from dry soils. Journal of Nematology 9, 176–181.

Fuchs, A.G., 1937. Neue parasitische und halbparasitische nematoden bei borkenkäfern und einige andere nematoden. I Teil. Zoologische Jahrbücher, Abteilung für Systematik. Ökologie und Geographie der Tiere 70, 291–380.

Handoo, Z.A., 1998. Plant-Parasitic Nematodes. http://www.ars.usda.gov/Services/docs.htm.

Heyns, J., 1962. *Elaphonema mirabile* n. gen., n. sp. (Rhabditida), a remarkable new nematode from South Africa. Proceedings of Helminthological Society of Washington 29, 128–130.

Heyns, J., 1963. A report of South African nematodes of the genera *Labronema* Thorne, *Discolaimus* Cobb, Discolaimoides n. gen. and *Discolaimium* (Nemata: Dorylaimoidea). Proceedings of the Helminthological Society of Washington 30, 7–15.

Heyns, J., 1965. On the Morphology and Taxonomy of the Aporcelaimidae, a New Family of Dorylaimoid Nematodes. Entomology Memoir, vol. 10. Department of Agricultural Technical Services, Republic of South Africa, pp. 1–51.

Heyns, J., 1968. A monographic study of the nematode families Nygolaimidae and Nygolaimellidae. Entomology Memoir Department of Agriculture, Republic of South Africa 19, 1–144.

Inglis, W.G., 1983. An outline classification of the phylum Nematoda. Australian Journal of Zoology 31, 243–255.

Jairajpuri, M.S., 1965. Three new species of the genus *Tylencholaimus* De Man, 1876 (Nematoda: Dorylaimoidea) from India. Nematologica 10, 515–518.

Jairajpuri, M.S., 1969. Studied on Monochida of India I. The genera *Hadronchus, Iotonchus* and *Miconchus* and revised classification of Monochida, new order. Nematologica 15, 557–581.

Jairajpuri, M.S., 1983. Observations on *Campydora* (Nematoda: Dorylaimida). Nematologia Mediterranea 11, 33–42.

Jairajpuri, M.S., Siddiqi, A.H., 1964. On a new nematode genus *Nordia* (Dorylaimoidea: Nordiinae n. subfam.) with remarks on the genus *Longidorella* Thorne, 1939. Proceedings of the Helminthological Society of Washington 31, 1–9.

Kanzaki, N., Giblin-Davis, R.M., Scheffrahn, R.H., Center, B.J., Davies, K.A., 2009. *Pseudaphelenchus yukiae* n. gen., n. sp. (Tylenchina: Aphelenchoididae) associated with *Cylindrotermes macrognathus* (Termitidae:Termitinae) in La Selva, Costa Rica. Nematology 11, 869–881.

Khan, S.H., 1973. On the proposal for *Neothada* n.gen. (Nematoda: Nothotylenchinae). 43rd Annual Session, Section B: Biological Sciences. Proceedings of the National Academy of Sciences, India 17–18.

Lorenzen, S., 1979. Marine Monhysteridae (sensu stricto, Nematodes) von der südchilenischen Küste und aus den küstenfernen Sublitoral der Nordsee. Studies on Neotropical Fauna and Environment 14, 203–214.

Micoletzky, H., 1922. Die freilebended Erd-Nematoden mit besonderer Bercksichtigung der Steiermark und der Bukowina, zugleich mit einer Revision samtlicher nichtmariner, freilebender Nematoden in Form Von Genus-Beschreibungen und Bestimmungsschlusseln. Archiv für Naturgeschichte 87, 1–650.

Micoletzky, H., 1925. Die freilebenden, Süsswasser-und Moomematoden Dänemarks. Nebst Anhang: über Amöbosporidien und andere Parasiten bei freilebenden Nematoden. Det Kongelige Danske Videnskabernes-Selskabs Skrivter 8, 57–310.

Mulk, M.M., Jairajpuri, M.S., 1974. Proposal of a new genus *Dolichorhynchus* and a new species *Dolichorhynchus nigericus* (Nematoda: Dolichorhynchidae). Indian Journal of Zoology 2, 15–18.

Nesterov, P.I., 1970. *Acromoldavicus* n. gen. and redescription of the species *Acrobeloides skrajavini* Nesterov & Lisethkaja, 1965 (Nematoda: Cephalobidae). In: Parasites of Animals and Plants, vol. 5. RIO Akademiya Nauk Moldavskoi SSR, Kishinev, pp. 134–138.

Nickle, W.R., 1970. A taxonomic review of the genera of the Aphelenchoidea (Fuchs, 1937) Thorne, 1949 (Nematoda: Tylenchida). Journal of Nematology 2, 375–392.

Nicoll, W., 1935. Rhabditida. Anguinidae. In: VI. Vermes. Zoological Record, 72, p. 105.

Örley, L., 1880. Az Anguillulidák magánrajza, (Monographie der Anguilluliden). Természetr Füzetek (Budapest) 4, 16–150.

Paramonov, A.A., 1967. A critical review of the suborder Tylenchina (Filipjev, 1934) (Nematoda: Secernentea). Trudy Gelmintologicheskoi Laboratorii Akademiya Nauk Sssr 18, 78–101 (in Russian).

Pearse, A.S., 1936. Zoological Names. A List of Phyla, Classes and Orders. Section F. Meeting of the American Association for the Advancement of Science. Duke University Press, Durham, North Carolina, USA. 24 pp.

Pearse, A.S., 1942. Introduction to Parasitology. Springfield, III, Baltimore, Maryland. 357 pp.

Pickup, J., Rothery, P., 1991. Water-loss and anhydrobiotic survival of nematodes of Antarctic fell-fields. Oikos 61, 379–388.

Platt, H.M., 1994. Foreword. In: Lorenzen, S. (Ed.), The Phylogenetic Systematics of Free-Living Nematodes. Ray Society, London. 383 pp.

Sasser, J.N., Freckman, D.W., 1987. A world perspective on nematology: the role of the society. In: Veech, J.A., Dickson, D.W. (Eds.), Vistas on Nematology: A Commemoration of the Twenty-Fifth Anniversary of the Society of Nematologists. Society of Nematologists, Hyattsville, USA, pp. 7–14.

Siddiqi, M.R., 1959. Basiria graminophila n. g., n. sp. (Nematoda: Tylenchinae) found associated with grass roots in Aligarh, India. Nematologica 4, 217–222.

Siddiqi, M.R., 1970. On the plant-parasitic nematode genera *Merlinius* gen. n. and *Tylenchorhynchus* Cobb and the classification of the families Dolichoridae and Belonolaimidae n. rank. Proceedings of the Helminthological Society 37, 68–77.

Siddiqi, M.R., 1971. On the plant-parasitic nematode genera *Histotylenchus* gen. n. and *Telotylenchoides* gen. n. (Telotylenchinae), with observations on the genus *Paratrophurus* Arias (Trophurinae). Annales Du Midi 17, 190–200.

Siddiqi, M.R., 1980. The origin and phylogeny of the nematode orders Tylenchida Thorne, 1949 and Aphelenchida n. ord. Helminthological Abstracts 49, 143–170.

Siddiqi, M.R., 1983. Phylogenetic relationships of the soil nematode orders Dorylaimida, Mononochida, Triplonchida and Alaimida, with a revised classification of the subclass Enoplia. Pakistan Journal of Nematology 1, 79–110.

Skarbilovich, T.S., 1947. Revision of the systematics of the nematode family Anguillulinidae Baylis and Daubney, 1926. Doklady Akademii nauk Ukrainskoi SSR 57, 307–308 (in Russian).

Steiner, G., 1936. Opuscula miscellanea nematologica, IV. Proceedings of the Helminthological Society 3, 74–80.

Steiner, G., 1945. *Helicotylenchus*, a new genus of plant-parasitic nematodes and its relationship to *Rotylenchus* Filipjev. Proceedings of the Helminthological Society of Washington 12, 34–38.

Steiner, G., 1949. Plant nematodes the grower should know. Proceedings Soil Science Society of Florida 4B, 72–117.

Taylor, A.L., 1936. The genera and species of the Criconematinae, a sub-family of the Anguillulinidae (Nematoda). Transactions of the American Microscopical Society 55, 391–421.

Thorne, G., 1930. Predacious nemas of the genus *Nygolaimus* and a new genus *Sectonema*. Journal of Agricultural Research 41, 445–466.

Thorne, G., 1935. Nemic parasites and associates of the mountain pine beetle (*Dendroctonus monticolae*) in Utah. Journal of Agricultural Research 51, 131–144.

Thorne, G., 1937. A revision of the nematode family Cephalobidae Chitwood and Chitwood, 1934. Proceedings of Helminthological Society of Washington 4, 1–16.

Thorne, G., 1938. Notes on free-living and plant-parasitic nematodes. IV. Proceedings of Helminthological Society of Washington 5, 64–65.

Thorne, G., 1939. Notes on free-living and plant-parasitic nematodes. V. Proceedings of Helminthological Society of Washington 6, 30–32.

Thorne, G., 1941. Some nematodes of the family Tylenchidae which do not possess a valvular median esophageal bulb. Great Basin Naturalist 2, 37–85.

Thorne, G., 1949. On the classification of the Tylenchida, new order (Nematoda, Phasmidia). Proceedings of Helminthological Society of Washington 16, 37–73.

Thorne, G., 1964. Nematodes of Puerto Rico: Belondiroidea, New Superfamily, Leptonchidae Thorne, 1935, and Belonenchidae New Family (Nematoda, Adenophprea, Dorylamida). Technical Paper University of Puerto Rico Agricultural Experiment Station No. 39. 51 pp.

Thorne, G., 1974. Nematodes of the Northern Great Plains. Part II. Dorylaimoidea in part (Nemata: Adenophorea). South Dakota State University Agriculture Experimental Station Technical Bulletin 41. 120 pp.

Thorne, G., Swanger, H.H., 1936. A monograph of the nematode genera *Dorylaimus* Dujardin, *Aporcelaimus* n.g., *Dorylaimoides* n.g. and *Pungentus* n.g. Capita Zoologica 6, 1–223.

von Linstow, O.F.B., 1877. Helminthologica. Archiv für Naturgeschichte 43, 1–18.

Whitehead, A.G., 1958. *Rotylenchoides brevis* n. g., n. sp. (Rotylenchoidinae n. subfam: Tylenchida). Nematologica 3, 327–331.

Womersley, C., Ching, C., 1989. Natural dehydration regimes as prerequisite for the successful induction of anhydrobiosis in the nematode *Rotylenchulus reniformis*. Journal of Experimental Biology 143, 359–372.

Yeates, G.W., 1967. Observation on phylogeny and evolution in the Dorylaiminae (Nematoda). New Zealand Journal of Science 10, 683–700.

Chapter 4

Advances and Perspectives in Soil Nematode Ecology in China

4.1 PROGRESS AND RESEARCH FIELD IN SOIL NEMATODE ECOLOGY

As awareness of the diversity and ecological significance of nematodes has increased, they have increasingly been used as indicators in the areas of biodiversity and sustainability in China. Emphasis has shifted from plant parasitic nematodes to free-living nematodes. Hence, soil nematode ecology has gradually developed into a discipline in itself that focuses on the relationship between soil nematode assemblage and its surrounding (abiotic and biotic) environment.

In China, research on soil free-living nematodes started in the 1990s (Yin, 1992, 1997). With the increase in interest in soil ecology, numerous researchers started using soil nematodes as bioindicators in farmland, forest, grassland, and wetland ecosystems to understand better the roles of soil biota and their influence on soil nutrient cycling and ecosystem functioning. Soil nematode ecology in China is booming in recent years (Fu et al., 2009). Search results on the *Web of Science* using "soil nematode" and "China" as key words showed that published articles have increased in past decades (Fig. 4.1). Therefore, we briefly review developments in soil nematode ecology in different ecosystems in China.

In farmland ecosystems, most research has been conducted on the influence of different management regimes on soil nematodes for a better understanding of their roles in nutrient cycling and plant growth. In a Hapli-Udic Cambosol of Northeast China, Liang et al. (2009) found that long-term application of nitrogen fertilizer and organic manure could change the availability of resources to the soil food web. The higher enrichment index and lower channel index suggested an enriched soil food web with bacterial decomposition channels dominant in the soil food web in the treatments with organic manure. Similar findings were also reported under greenhouse conditions, in which the relative abundance of bacterivores increased and that of plant parasites decreased after long-term manure applications (Li et al., 2010). In a wheat–maize rotation system, Zhang et al. (2016a) found that organic management (manure or straw) combined with nitrogen–phosphorus–potassium fertilizer could effectively enhance the association between microbial and nematode communities. Both manure application and the incorporation of straw increased nematode

229

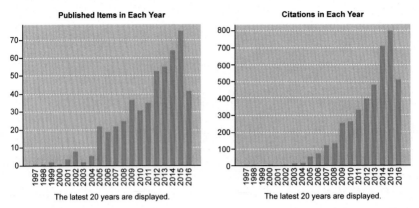

FIGURE 4.1 Scientific articles published on the soil nematode community in China (1997–2016).

functional metabolic footprints within all aggregates. The incorporation of crop straw effectively improved soil physicochemical properties and benefitted nematode survival within small aggregate size fractions (Zhang et al., 2016b).

In the black soil (Typic Hapludoll, USDA Soil Taxonomy) of Northeast China, Zhang et al. (2015a,b) found that 10 years of conservation tillage could effectively enhance the structure and function of soil food webs; higher abundances of protozoa, bacterivorous, and omnivorous-carnivorous nematodes were found in no tillage (NT) and ridge tillage (RT) compared with conventional tillage systems. Among the four aggregate fractions, soil microbial biomass and diversity were higher in microaggregates (<0.25 mm), whereas soil nematode abundance and diversity were higher in large macroaggregates (>2 mm). The increase in microbial biomass and nematode abundance at the micro-niche within aggregates could contribute to the relatively higher C sequestration in conservation tillage systems (NT and RT) (Zhang et al., 2013).

Using the microcosm experiment, Li and Hu (2001) quantified the effect of bacterivorous nematodes on plant growth and nutrient absorption, and found that bacterivorous nematodes stimulated the growth of wheat and the uptake of nitrogen. The presence of bacterivorous nematodes led to greater NH_4^+ and NO_3^- content, which indicated that bacterivorous nematodes promoted both N mineralization and nitrification (Xiao et al., 2010). The effects of bacteria and nematodes on nutrient and hormone concentrations were responsible for the increase in plant growth (Mao et al., 2007; Xu et al., 2015).

In farmland ecosystems, some researchers also studied the response of soil nematodes to climate change and their feedback to plant growth. Using the microcosm experiment, Luo et al. (2008) studied the influence of nematodes and earthworms on the emissions of soil trace gases (CO_2 and N_2O) and found that soil with greater nematode populations had higher CO_2 and N_2O emissions.

In the free-air O_3 enrichment platform, Li et al. (2012) observed that the responses of soil biota to elevated O_3 were greater in ozone-tolerant wheat than in ozone-sensitive wheat. Importantly, this effect of elevated ozone remained present after the ozone treatment had ceased and was able to influence plant growth and soil characteristics further via legacy effects on soil biota (Li et al., 2015, 2016).

In forest ecosystems, more emphasis was put on the effects of vegetation, forest type, and forest management practices on soil nematodes. In a subtropical evergreen broad-leaved forest in Southwest China, girdling reduced total nematode density in the humus layer. The reduced density of fungivores in the girdled plots suggested that girdling induced a modified energy flow through the fungal-based pathway (Li et al., 2009). In a subtropical forest, Shao et al. (2016) studied the effect of plant removal on the soil micro-food web and found that plant removal decreased nematode richness and diversity, and increased nematode dominance. They concluded that maintaining plant diversity was important for the complexity and stability of soil food webs and the sustainable management of subtropical bamboo forests. In Xishuangbanna, Xiao et al. (2014) showed that natural forests had the highest nematode abundance and generic richness. The rubber monoculture had reduced soil C and N and nematode generic richness compared with the natural forest, indicating the loss of soil nutrients and ecological functioning after change in land use. In the Changbai Mountain of northeast China, Zhang et al. (2012) reported that soil nematode abundance and diversity differed significantly among different forest types along an elevation gradient from 760 to 1950 m. In forests of different ages (young, middle, and old), Zhang et al. (2015a,b) observed a lower enrichment footprint in the young forest and a higher enrichment footprint in the middle-aged forest, indicating an increase in resource entry into the soil food web from the young to the middle-aged forest of the Changbai Mountain. In addition, nutrient availability could also regulate responses in forest ecosystems. The abundance of fungivores, plant parasites, and omnivores-predators was reduced by water combined with added N, and the impacts of N deposition on soil nematode communities were mediated by water availability in the temperate forest (Sun et al., 2013). In a tropical forest, Zhao et al. (2014) found that the addition of nitrogen and/or phosphorus suppressed the abundance of soil nematodes. Moreover, the addition of phosphorus decreased trophic linkages and reduced the use of carbon by the soil nematode community in the secondary tropical forest, leading to a more fungal-dominated decomposition pathway.

In grassland ecosystems, research on nematode ecology is limited compared with that in agro-ecosystems. Current research on grasslands mainly focuses on soil nematodes' responses to climatic change, grassland restoration, and management practices. In the grassland of Inner Mongolia, heavy grazing has significantly influenced soil microbial biomass, the abundance of protozoa, and soil nematodes (Chen et al., 2013; Qi et al., 2011). Whereas the application of small amounts of nitrogen to N-limited semiarid grasslands did not induce changes in short-term research conducted by Ruan et al. (2012). In the multilevel N addition experiments, a high level of nitrogen addition was found to affect soil microbial community and nematode composition negatively through soil acidification (Chen et al., 2015; Wei et al., 2013). In the grassland ecosystems of the Songnen Plain, the abundance and diversity of the soil nematode community were influenced by the restoration of vegetation (Wu et al., 2008). In alpine meadow ecosystems, continuous grazing greatly changed the composition and diversity of soil nematodes, indicating a radical shift in biodiversity and belowground processes in the Tibetan plateau grassland (Hu et al., 2015).

In wetland ecosystems, research mainly centered on the effects of land use management and exotic plant invasion on soil nematode communities (Chen et al., 2007; Wu et al., 2002). On Chongming Island of the Yangtze River estuary, Wu et al. (2005) found that marsh reclamation and its frequency substantially altered the nematode community structure, with lower generic richness and diversity observed at reclaimed stations than at tideland stations. In the Yangtze Delta region, the exotic plant *Spartina alterniflora* stimulated the growth of bacterivorous nematodes by producing a higher quality of litter than the native *Phragmites australis*. This indicated that the invasion of exotic plants might alter the functions of an ecosystem through their effects on soil decomposer communities (Chen et al., 2007).

In the sand dune ecosystems, research focused mostly on desertification management practices and the formation of biological soil crusts on soil nematode communities (Guan et al., 2015; Zhang et al., 2010). In the Tengger Desert, Liu et al. (2011) studied the effects of biological soil crusts on soil nematode communities after dune stabilization. They found that the colonization and development of biological soil crusts enhanced soil nematode diversity. In the Horqin Desert, Guan et al. (2015) found that the conversion of drifting sand dunes to shrub land resulted in significant changes in the proportion of r- (cp-2) and K-strategists (cp-4) of soil nematode communities. The variation in soil nematode communities could reflect the restoration process of the sandy ecosystems after the establishment of *Caragana microphylla*.

In short, nematode ecology research in China has covered different ecosystems and is prospering. With the development of soil ecology and the increase in interest for aboveground and belowground interactions, it is necessary to monitor the network of soil biota across different ecosystems and different areas (such as transects from north to south and from west to east); it will help us gain a more comprehensive knowledge of soil biodiversity in China. Fortunately, some Chinese scholars have begun to devote themselves to this kind of work with the support of the Chinese Academy of Sciences. A conference on a soil fauna diversity monitoring network was held in 2016, and scholars have started to work together to establish a Chinese soil fauna monitoring network and to provide a fundamental database for the protection of biodiversity and conservation of resources.

4.2 FUTURE PERSPECTIVES OF SOIL NEMATODE ECOLOGY IN CHINA

For a better understanding of soil biodiversity and the role it has in belowground subsystems, on the one hand, we need to train and support more taxonomic experts to discover and understand more about biological diversity. On the other hand, we still need to expand research in several aspects.

4.2.1 Connection of Soil Nematodes With Other Soil Biota Within Food Webs

Research has emphasized the potential role of the soil food web in nutrient cycling and carbon sequestration (De Vries et al., 2013; De Vries and Caruso, 2016; Kardol et al., 2016; Wall et al., 2015). Interactions between soil nematodes and other biota within food webs could affect the structure and function of the soil ecosystem (Wurst et al., 2008). To date, we still lack a clear understanding of how soil nematodes and other soil biota within food webs contribute to the processes and services of the ecosystem, especially for C cycling (Filser et al., 2016). The relatively few studies that exist quantitatively characterize C dynamically at the level of entire soil food web (Pausch et al., 2016). Thus, understanding the roles of the soil food web in regulating belowground processes, such as decomposition, nutrient cycling, and C cycling, is a hot topic in soil ecological research (Shaw et al., 2016).

Stable isotope technologies offer a means to follow and quantify C in studies on terrestrial nutrient dynamics in situ (Ostle et al., 2000; Pausch et al., 2016; Shaw et al., 2016). The flow of C from plants and soil organic matter into the soil food web is initiated by organisms at the entry level (e.g., fungi, bacteria, and plant-feeding nematodes) and then at the trophic level (e.g., protozoa, some nematodes, and mesofauna) (Minoshima et al., 2007). The lipid pattern in soil biota has been used successfully to assess trophic interactions either solely by fatty acid profiling or combined with stable isotope techniques, i.e., compound-specific analysis of the $^{13}C-^{12}C$ ratio in individual fatty acids (Ruess and Chamberlain, 2010). However, pulse-labeling studies tracing the ^{13}C signal in fatty acids at higher trophic levels of the soil food chain are lacking. Therefore, research on connections among soil biota in food webs using new technology needs to be strengthened.

4.2.2 Feedback Between Aboveground and Belowground Communities

Aboveground plant diversity and identity are key forces of belowground communities and are highly relevant to the diversity and composition of the soil biotic community (Bezemer et al., 2010; De Deyn et al., 2004). Changes in soil biotic and abiotic characteristics may directly or indirectly influence plant growth (van der Heijden et al., 2008). This aboveground–belowground feedback is called plant–soil feedback effects (van der Putten et al., 2013). Understanding the linkages between aboveground and belowground communities has emerged as an important challenge given that soil biota are not only a "black box" of highly redundant species, they have a range of ecosystem functions (Milcu et al., 2013; Wardle et al., 2004). To date, our research on linkages between aboveground and belowground has been limited in China. Understanding the relations between plants and soil organisms is essential to predicting the consequences of environmental change on the processes and functioning of ecosystems.

4.2.3 Molecular Analysis of Soil Nematode Diversity

To obtain a complete picture of the highly complex soil community, some scholars have started to use molecular techniques to analyze soil nematode diversity (Wu et al., 2009). DNA-based methods used to assess nematode assemblage structure and diversity are limited in China. Thus, molecular methods that rapidly assess the "species"-level diversity of soil nematodes could potentially be an important tool in the study of soil ecology (Kerfahi et al., 2016; Yin et al., 2010). However, some molecular methods mainly use relative abundance to calculate the biodiversity of soil animals; they may not measure the density, absolute abundance, or biomass of individual soil nematodes. Molecular sequencing alone on a large scale will not improve our understanding of the important role of soil biodiversity. Until such methods are authenticated, we have to follow traditional methods for identifying nematodes at a generic or family level. In the future, standard monitoring approaches should be established to combine traditional identification methods with molecular analysis techniques to enhance our understanding of the functioning and services provided by soil organisms. An integration of biodiversity at different levels from functional groups to genetic differences (Kardol et al., 2016) will enhance our understanding of the importance of soil biodiversity in driving ecosystem processes and provide useful knowledge to land managers and decision makers.

REFERENCES

Bezemer, T.M., Fountain, M., Barea, J., Christensen, S., Dekker, S., Duyts, H., Van Hal, R., Harvey, J.A., Hedlund, K., Maraun, M., Mikola, J., Mladenov, A.G., Robin, C., De Ruiter, P., Scheu, S., Setälä, H., Šmilauer, P., van der Putten, W.H., 2010. Divergent composition but similar function of soil food webs of individual plants: plant species and community effects. Ecology 91, 3027–3036.

Chen, D.M., Zheng, S.X., Shan, Y.M., Taube, F., Bai, Y.F., 2013. Vertebrate herbivore-induced changes in plants and soils: linkages to ecosystem functioning in a semi-arid steppe. Functional Ecology 27, 273–281.

Chen, D.M., Lan, Z.C., Hu, S.J., Bai, Y.F., 2015. Effects of nitrogen enrichment on belowground communities in grassland: relative role of soil nitrogen availability vs. soil acidification. Soil Biology & Biochemistry 89, 99–108.

Chen, H.L., Li, B., Fang, C.M., Chen, J.K., Wu, J.H., 2007. Exotic plant influences soil nematode communities through litter input. Soil Biology & Biochemistry 39, 1782–1793.

De Deyn, G.B., Raaijmakers, C.E., van Ruijven, J., Berendse, F., van der Putten, W.H., 2004. Plant species identity and diversity effects on different trophic levels of nematodes in the soil food web. Oikos 106, 576–586.

De Vries, F.T., Caruso, T., 2016. Eating from the same plate? Revisiting the role of labile carbon inputs in the soil food web. Soil Biology and Biochemistry 102, 4–9.

De Vries, F.T., Thébault, E., Liiri, M., Birkhofer, K., Tsiafouli, M.A., Bjørnlund, L., Jørgensen, H.B., Brady, M.V., Christensen, S., De Ruiter, P.C., Hertefeldt, T., Frouz, J., Hedlund, K., Hemerik, L., Hol, W.H.G., Hotes, S., Mortimer, S.R., Setälä, H., Sgardelis, S.P., Uteseny, K., van der Putten, W.H., Wolter, V., Bardgett, R.D., 2013. Soil food web properties explain ecosystem services across European land use systems. Proceedings of the National Academy of Sciences of the United States of America 110, 14296–14301.

Filser, J., Faber, J.H., Tiunov, A.V., Brussaard, L., Frouz, J., De Deyn, G., Uvarov, A.V., Berg, M.P., Lavelle, P., Loreau, M., Wall, D.H., Querner, P., Eijsackers, H., Jiménez, J.J., 2016. Soil fauna: key to new carbon models. Soil 2, 565–582.

Fu, S.L., Zou, X.M., Coleman, D.C., 2009. Highlights and perspectives of soil biology and ecology research in China. Soil Biology & Biochemistry 41, 868–876.

Guan, P.T., Zhang, X.K., Yu, J., Ma, N.N., Liang, W.J., 2015. Variation of soil nematode community composition with increasing sand-fixation year of Caragana microphylla: bioindication for desertification restoration. Ecological Engineering 81, 93–101.

Hu, J., Wu, J.H., Ma, M.J., Nielsen, U.N., Wang, J., Du, G.Z., 2015. Nematode communities response to long-term grazing disturbance on Tibetan plateau. European Journal of Soil Biology 69, 24–32.

Kardol, P., Throop, H.L., Adkins, J., de Graaff, M.A., 2016. A hierarchical framework for studying the role of biodiversity in soil food web processes and ecosystem services. Soil Biology & Biochemistry 102, 33–36.

Kerfahi, D., Tripathi, B.M., Porazinska, D.L., Park, J., Go, R., Adams, J.M., 2016. Do tropical rain forest soils have greater nematode diversity than high arctic tundra? A metagenetic comparison of Malaysia and Svalbard. Global Ecology and Biogeography 25 (6), 716–728.

Li, H.X., Hu, F., 2001. Effects of bacterial-feeding nematode inoculation on wheat growth and N and P uptake. Pedosphere 11, 57–62.

Li, Q., Jiang, Y., Liang, W.J., Lou, Y.L., Zhang, E.P., Liang, C.H., 2010. Long-term effect of fertility management on the soil nematode community in vegetable production under greenhouse conditions. Applied Soil Ecology 46 (1), 111–118.

Li, Q., Bao, X.L., Lu, C.Y., Zhang, X.K., Zhu, J.G., Jiang, Y., Liang, W.J., 2012. Soil microbial food web responses to free air ozone enrichment can depend on the ozone-tolerance of wheat cultivars. Soil Biology & Biochemistry 47, 27–35.

Li, Q., Yang, Y., Bao, X.L., Liu, F., Liang, W.J., Zhu, J.G., Bezemer, T.M., van der Putten, W.H., 2015. Legacy effects of elevated ozone on soil biota and plant growth. Soil Biology & Biochemistry 91, 50–57.

Li, Q., Yang, Y., Bao, X.L., Zhu, J.G., Liang, W.J., Bezemer, T.M., 2016. Cultivar specific plant-soil feedback overrules soil legacy effects of elevated ozone in a rice-wheat rotation system. Agriculture, Ecosystems and Environment 232, 85–92.

Li, Y.J., Yang, X., Zou, X.M., Wu, J.H., 2009. Response of soil nematode communities to tree girdling in a subtropical evergreen broad-leaved forest of southwest China. Soil Biology & Biochemistry 41, 877–882.

Liang, W.J., Lou, Y.L., Li, Q., Zhong, S., Zhang, X.K., Wang, J.K., 2009. Nematode faunal response to long-term application of nitrogen fertilizer and organic manure in Northeast China. Soil Biology & Biochemistry 41 (5), 883–890.

Liu, Y.M., Li, X.R., Jia, R.L., Huang, L., Zhou, Y.Y., Gao, Y.H., 2011. Effects of biological soil crusts on soil nematode communities following dune stabilization in the Tengger Desert, Northern China. Applied Soil Ecology 49, 118–124.

Luo, T.X., Li, H.X., Wang, T., Hu, F., 2008. Influence of nematodes and earthworms on the emissions of soil trace gases (CO_2, N_2O). Acta Ecologica Sinica 28, 993–999 (in Chinese).

Mao, X.F., Hu, F., Griffiths, B., Chen, X.Y., Liu, M.Q., Li, H.X., 2007. Do bacterial-feeding nematodes stimulate root proliferation through hormonal effects? Soil Biology & Biochemistry 39, 1816–1819.

Milcu, A., Allan, E., Roscher, C., Jenkins, T., Meyer, S.T., Flynn, D., Bessler, H., Buscot, F., Engels, C., Gubsch, M., König, S., Lipowsky, A., Loranger, J., Renker, C., Scherber, C., Schmid, B., Thébault, E., Wubet, T., Weisser, W.W., Scheu, S., Eisenhauer, N., 2013. Functionally and phylogenetically diverse plant communities key to soil biota. Ecology 94, 1878–1885.

Minoshima, H., Jakson, L.E., Cavagnaro, T.R., Sánchez-Moreno, S., Ferris, H., Temple, S.R., Goyal, S., Mitchell, J.P., 2007. Soil food webs and carbon dynamics in response to conservation tillage in California. Soil Science Society of America Journal 71, 952–963.

Ostle, N., Ineson, P., Benham, D., Sleep, D., 2000. Carbon assimilation and turnover in grassland vegetation using an in situ $^{13}CO_2$ pulse labelling system. Rapid Communiactions in Mass Spectrometry 14, 1345–1350.

Pausch, J., Kramer, S., Scharroba, A., Scheunemann, N., Butenschoen, O., Kandeler, E., Marhan, S., Riederer, M., Scheu, S., Kuzyakov, Y., Ruess, L., 2016. Small but active – pool size does not matter for carbon incorporation in below-ground food webs. Functional Ecology 30, 479–489.

Qi, S., Zheng, H.X., Lin, Q.M., Li, G.T., Xi, Z.H., Zhao, X.R., 2011. Effects of livestock grazing intensity on soil biota in a semiarid steppe of Inner Mongolia. Plant and Soil 340, 117–126.

Ruan, W.B., Sang, Y., Chen, Q., Zhu, X., Lin, S., Gao, Y.B., 2012. The response of soil nematode community to nitrogen, water, and grazing history in the Inner Mongolian Steppe, China. Ecosystems 15, 1121–1133.

Ruess, L., Chamberlain, P.M., 2010. The fat that matters: soil food web analysis using fatty acids and their carbon stable isotope signature. Soil Biology & Biochemistry 42, 1898–1910.

Shao, Y.H., Wang, X.L., Zhao, J., Wu, J.P., Zhang, W.X., Neher, D.A., Li, Y.X., Lou, Y.P., Fu, S.L., 2016. Subordinate plants sustain the complexity and stability of soil micro-food webs in natural bamboo forest ecosystems. Journal of Applied Ecology 53, 130–139.

Shaw, E.A., Denef, K., De Tomasel, C.M., Cotrufo, M.F., Wall, D.H., 2016. Fire affects root decomposition, soil food web structure, and carbon flow in tallgrass prairie. Soil 2, 199–210.

Sun, X.M., Zhang, X.K., Zhang, S.X., Han, S.J., Dai, G.H., Liang, W.J., 2013. Soil nematode responses to increases in nitrogen deposition and precipitation in a temperate forest. PLoS One e82468.

van der Heijden, M.G., Bardgett, R.D., Van Straalen, N.M., 2008. The unseen majority: soil microbes as drivers of plant diversity and productivity in terrestrial ecosystems. Ecology Letters 11, 296–310.

van der Putten, W.H., Bardgett, R.D., Bever, J.D., Bezemer, T.M., Casper, B.B., Fukami, T., Kardol, P., Klironomos, J.N., Kulmatiski, A., Schweitzer, J.A., Suding, K.N., Van de Voorde, T.F.J., Wardle, D.A., 2013. Plant-soil feedback: the past, the present and future challenges. Journal of Ecology 101, 265–276.

Wall, D.H., Nielsen, U.N., Six, J., 2015. Soil biodiversity and human health. Nature 528, 69–76.

Wardle, D.A., Bardgett, R.D., Klironomos, J.N., Setälä, H., van Der Putten, W.H., Wall, D.H., 2004. Ecological linkages between aboveground and belowground biota. Science 304, 1629.

Wei, C.Z., Yu, Q., Bai, E., Lü, X.T., Li, Q., Xia, J.Y., Kardol, P., Liang, W.J., Wang, Z.W., Han, X.G., 2013. Nitrogen deposition weakens plant–microbe interactions in grassland ecosystems. Global Change Biology 19, 3688–3697.

Wu, D.H., Yin, W.Y., Pu, Z.Y., 2008. Changes among soil nematode community characteristics in relation to different vegetation restoration practices. Acta Ecologica Sinica 28, 1–12 (in Chinese).

Wu, T., Ayres, E., Li, G., Bardgett, R.D., Wall, D.H., Garey, J.R., 2009. Molecular profiling of soil animal diversity in natural ecosystems: incongruence of molecular and morphological results. Soil Biology & Biochemistry 41, 849–857.

Wu, J.H., Fu, C.Z., Chen, S.S., Chen, J.K., 2002. Soil faunal response to land use: effect of estuarine tideland reclamation on nematode communities. Applied Soil Ecology 21, 131–147.

Wu, J.H., Fu, C.Z., Lu, F., Chen, J.K., 2005. Changes in free-living nematode community structure in relation to progressive land reclamation at an intertidal marsh. Applied Soil Ecology 29, 47–58.

Wurst, S., Allema, B., Duyts, H., van der Putten, W.H., 2008. Earthworms counterbalance the negative effects of microorganisms on plant diversity and enhance the tolerance of grasses to nematodes. Oikos 117, 711–718.

Xiao, H.F., Griffiths, B., Chen, X.Y., Liu, M.Q., Jiao, J.G., Hu, F., Li, H.X., 2010. Influence of bacterial-feeding nematodes on nitrification and the ammonia-oxidizing bacteria (AOB) community composition. Applied Soil Ecology 45, 131–137.

Xiao, H.F., Tian, Y.H., Zhou, H.P., Ai, X.S., Yang, X.D., Schaefer, D.A., 2014. Intensive rubber cultivation degrades soil nematode communities in Xishuangbanna, southwest China. Soil Biology & Biochemistry 76, 161–169.

Xu, L., Xu, W.S., Jiang, Y., Hu, F., Li, H.X., 2015. Effects of interactions of auxin-producing bacteria and bacterial-feeding nematodes on regulation of peanut growths. PLoS One 10, e0124361.

Yin, X.Q., Song, B., Dong, W.H., Xin, W.D., 2010. A review on the eco-geography of soil fauna in China. Acta Geographica Sinica 65, 91–102 (in Chinese).

Yin, W.Y. (Ed.), 1992. Subtropical Soil Animals of China. Science Press, Beijing (in Chinese).

Yin, W.Y., 1997. Studies on soil animals in subtropical China. Agriculture, Ecosystems and Environment 62, 119–126.

Zhang, M., Liang, W.J., Zhang, X.K., 2012. Soil nematode abundance and diversity under different forest types in Changbai Mountain, China. Zoological Studies 51, 619–626.

Zhang, S.X., Li, Q., Lü, Y., Sun, X.M., Jia, S.X., Zhang, X.P., Liang, W.J., 2015a. Conservation tillage positively influences the microflora and microfauna in the black soil of Northeast China. Soil & Tillage Research 149, 46–52.

Zhang, S.X., Li, Q., Lü, Y., Zhang, X.P., Liang, W.J., 2013. Contributions of soil biota to C sequestration varied with aggregate fractions under different tillage systems. Soil Biology & Biochemistry 62, 147–156.

Zhang, X.K., Dong, X.W., Liang, W.J., 2010. Spatial distribution of soil nematode communities in stable and active sand dunes of Horqin Sandy Land. Arid Land and Research Management 24, 68–80.

Zhang, X.K., Guan, P.T., Wang, Y.L., Li, Q., Zhang, S.X., Zhang, Z.Y., Bezemer, T.M., Liang, W.J., 2015b. Community composition, diversity and metabolic footprints of soil nematodes in differently-aged temperate forests. Soil Biology & Biochemistry 80, 118–126.

Zhang, Z.Y., Zhang, X.K., Mahamood, Md., Zhang, S.Q., Huang, S.M., Liang, W.J., 2016a. Effect of long-term combined application of organic and inorganic fertilizers on soil nematode communities within aggregates. Scientific Reports 6, 31118.

Zhang, Z.Y., Zhang, X.K., Xu, M.G., Zhang, S.Q., Huang, S.M., Liang, W.J., 2016b. Responses of soil micro-food web to long-term fertilization in a wheat-maize rotation system. Applied Soil Ecology 98, 56–64.

Zhao, J., Wang, F.M., Li, J., Zou, B., Wang, X.L., Li, Z.A., Fu, S.L., 2014. Effects of experimental nitrogen and/or phosphorus additions on soil nematode communities in a secondary tropical forest. Soil Biology & Biochemistry 75, 1–10.

Index

'Note: Page numbers followed by "f" indicate figures, "t" indicate tables.'

A

Aboveground plant diversity, 233–234
Alaimida, 221
  Alaimina, 221
    Alaimoidea. *See* Alaimoidea
Alaimidae, 68
Alaimoidea
  Alaimidae, 222
    Alaimus, 223, 223f–224f
Anguinidae, 65–66
Anguinoidea, 137–138
  Anguinidae, 138
    Ditylenchus, 139–140, 139f–140f
Aphelenchi, 147
  Aphelenchina, 147
    Aphelenchoidea. *See* Aphelenchoidea
Aphelenchidae, 54, 55f, 66
Aphelenchoidea, 147–148
  Aphelenchidae, 148
    Aphelenchus, 148–149, 149f–150f
    Paraphelenchus, 150–151, 151f–152f
  Aphelenchoididae, 151–153
    Aphelenchoides, 153, 154f–155f
    Bursaphelenchus, 155–156,
      158f–159f
    Pseudaphelenchus, 154, 156f–157f
Aphelenchoididae, 54, 55f, 66
Aporcelaimidae, 54, 55f, 66
Araeolaimida, 92
  Leptolaimina, 92
    Haliplectoidea. *See* Haliplectoidea
    Plectoidea. *See* Plectoidea

B

Belondiridae, 67
Belondiroidea, 190–191
  Belondiridae, 191–192
    Axonchium, 192–193, 193f–194f
    Dorylaimellus, 193–194, 195f
Belowground biotic communities, 2
Belowground plant diversity, 233–234

C

Campydoridae, 54, 55f, 67
Campydoroidea, 209
  Campydoridae, 210
    Campydora, 210–211, 210f–211f
Cephalobidae, 14–17, 54, 55f, 64–68
Cephaloboidea, 69
  Cephalobidae, 70
    Acrobeles, 71–72, 71f–72f
    Acrobeloides, 72–73, 73f–74f
    Acromoldavicus, 73–74, 75f
    Cephalobus, 76–77, 76f–77f
    Cervidellus, 77–78, 78f–79f
    Chiloplacus, 78–79, 80f–81f
    Eucephalobus, 79–80, 82f–83f
Constrained ordination analysis, 39, 40t
Criconematidae, 54, 55f, 66
Criconematoidea, 141
  Criconematidae, 142
    Macroposthonia, 143–144,
      143f–144f

D

Digestive tract/system
  anus/cloaca, 57
  intestine, 56
  mesenteron, 51
  prerectum, 56–57
  proctodeum, 51
  rectum, 57
  stomodeum, 51
    classification, 51
    esophageal/pharyngeal glands, 54,
      55f–56f
    esophago-intestinal junction/cardia, 54
    stoma, 51–54, 52f–54f
Diphtherophoridae, 67
Diphtherophoroidea, 212–213
  Diphtherophora, 213, 214f–215f
Diplogastridae, 54, 55f
Diploscapteridae, 54, 55f

Diversity and distribution, 32f–33f
  aboveground vegetation and soil
    conditions, 39
  beta diversity, 36
  constrained ordination analysis, 39, 40t
  desert and desert steppe
    ecosystems, 14, 15f
  enrichment index (EI), 29, 36, 37f–38f
  generic richness and dominance, 29, 30f–31f
  mean annual precipitation (MAP), 14
  mean annual temperature (MAT), 14
  nematode channel ratio, 29, 34f–35f
  nematode faunal analysis, 36
  plant parameters, 39–40
  Shannon–Weaver diversity, 29
  structure index (SI), 29, 36, 37f–38f
  terrestrial ecosystems, 13
  trophic diversity, 29, 34f–35f
  typical and meadow steppe
    ecosystems, 14, 15f
  unconstrained gradient analysis, 17–18
Dolichodoridae, 14–17, 54, 55f, 65
Dolichodoroidea, 111–113
  Dolichodoridae, 113–114
    Dolichorhynchus, 117–118, 119f–120f
    Merlinius, 114, 115f–116f
    Quinisulcius, 119–120, 121f–122f
    Tylenchorhynchus, 115–117, 117f–118f
  Psilenchidae, 120–122
    Atetylenchus, 125–126, 125f–126f
    Psilenchus, 122–123, 123f–124f
Dorylaimida, 156–158
  Campydoryna, 209
    Campydoroidea. *See* Campydoroidea
  Dorylaimina, 158–159
    Belondiroidea. *See* Belondiroidea
    Dorylaimoidea. *See* Dorylaimoidea
    Longidoroidea. *See* Longidoroidea
    Tylencholaimoidea. *See*
      Tylencholaimoidea
  Nygolaimina, 205
    Nygolaimoidea. *See* Nygolaimoidea
Dorylaimidae, 54, 55f, 66
Dorylaim nematodes, 52–53, 53f
Dorylaimoidea, 160
  Aporcelaimidae, 163–165
    Aporcelaimellus, 167, 168f
    Aporcelaimus, 165–167, 166f
    Makatinus, 169, 169f–170f
    Sectonema, 170–171, 171f–172f
  Dorylaimidae, 160–161
    Mesodorylaimus, 161, 162f–163f
    Prodorylaimus, 161–162, 164f–165f

Nordiidae, 180–183
  Enchodelus, 185–186, 185f–186f
  Pungentus, 183–185, 184f
Qudsianematidae, 171–173
  Crassolabium, 180, 182f
  Discolaimus, 178–180, 179f
  Ecumenicus, 173, 174f–175f
  Epidorylaimus, 173–175, 176f
  Eudorylaimus, 175–178, 177f–178f
  Kochinema, 180, 181f

E

Enoplida, 87–88
  Tripyloidea. *See* Tripyloidea
Eurasian Steppe, 1

F

Field survey, grassland transect
  brown calcic and chestnut soil, 6–8
  brown-gray desert soil, 5–6
  desert ecosystems, 5, 6f
  desert steppes, 5–6, 6f–7f
  meadow steppe, 8, 8f
  mean annual temperature (MAT), 4–5
Food webs, 233
Forest ecosystems, 231

G

Genera and species
  Alaimida. *See* Alaimida
  Alaimidae, 68
  Anguinidae, 65–66
  Aphelenchidae, 66
  Aphelenchoididae, 66
  Aporcelaimidae, 66
  Araeolaimida. *See* Araeolaimida
  Belondiridae, 67
  Campydoridae, 67
  Cephalobidae, 64–68
  Criconematidae, 66
  Diphtherophoridae, 67
  Dolichodoridae, 65
  Dorylaimida. *See* Dorylaimida
  Dorylaimidae, 66
  Enoplida. *See* Enoplida
  Hoplolaimidae, 65
  Leptonchidae, 67
  Longidoridae, 67
  Monhysterida. *See* Monhysterida
  Monhysteridae, 64
  Mononchida. *See* Mononchida

Mydonomidae, 67
Mylonchulidae, 68
Nordiidae, 66–67
Nygolaimidae, 67
Panagrolaimidae, 64
Paratylenchidae, 66
Plectidae, 65
Pratylenchidae, 65
Prismatolaimidae, 67–68
Psilenchidae, 65
Qudsianematidae, 66
Rhabditida. *See* Rhabditida
Rhabdolaimidae, 64
structure
    amphids, 51
    body openings, 49
    body/somatic setae, 49
    cephalic framework, 50f, 51
    cuticular ornamentations, 49, 49f
    deirids, 51
    digestive tract/system. *See* Digestive
        tract/system
    hypodermis, 49–50
    lips and labial papillae, 50, 50f
    longitudinal markings, 48–49, 49f
    molting, 47–48
    organization, 47
    phasmids, 51
    posture, 47
    regions, 47, 48f
    reproductive system. *See* Reproductive
        system
    somatic musculature, 50
    symmetry, 47
    transverse markings, 48, 49f
Triplonchida. *See* Triplonchida
Tripylidae, 64
Tylenchida. *See* Tylenchida
Tylenchidae, 65
Tylencholaimidae, 67
Xiphinematidae, 67
Grassland, class and subclass, 1–2, 3t
Grassland ecosystems, 231–232

H

Haliplectoidea, 92–93
    Rhabdolaimidae, 93
        Udonchus, 93–94, 94f–95f
Hoplolaimidae, 54, 55f, 65
Hoplolaimoidea, 126–127
    Hoplolaimidae, 127–128
        Helicotylenchus, 131–133, 132f

Rotylenchoides, 133–134, 133f–134f
Rotylenchus, 129–131, 130f–131f
Scutellonema, 128, 129f–130f
Pratylenchidae, 134–135
    Pratylenchus, 135–137, 136f–137f

L

Leptonchidae, 67
Longidoridae, 67
Longidoroidea, 186–187
    Longidoridae, 187
        Longidorus, 187, 188f
    Xiphinematidae, 189
        Xiphinema, 189–190, 190f–191f

M

Monhysterida, 84–85
    Monhysterina, 85–86
        Monhysteroidea. *See* Monhysteroidea
Monhysteridae, 64
Monhysteroidea, 86
    Monhysteridae, 86–87
        Geomonhystera, 87, 88f–89f
Mononchida, 218
    Mononchina, 218
        Mononchoidea. *See* Mononchoidea
Mononchidae, 54, 55f
Mononchoidea, 218–219
    Mylonchulidae, 219
        Mylonchulus, 219, 220f
Mydonomidae, 67
Mylonchulidae, 68

N

Nematode faunal analysis, 36
Neodiplogastridae, 54, 55f
Nordiidae, 66–67
Nygolaimidae, 67
Nygolaimoidea, 205
    Nygolaimidae, 205–206
        Clavicaudoides, 206–207, 208f–209f
        Nygolaimus, 206, 207f–208f

P

Panagrolaimidae, 54, 55f, 64, 81–83
Panagrolaimoidea, 80–81
    Panagrolaimidae, 81–83
        Panagrellus, 83–84, 84f–85f
Paraphelenchidae, 54, 55f
Paratylenchidae, 54, 55f, 66

Plant–soil feedback effects, 233–234
Plectidae, 54, 55f, 65
Plectoidea, 94
  Plectidae, 95–96
    Plectus, 96, 97f–98f
    Wilsonema, 96–97, 99f–100f
Pratylenchidae, 54, 55f, 65
Principal component analysis
    (PCA), 17–18, 18f
Prismatolaimidae, 67–68
Prismatolaimoidea, 214–215
  Prismatolaimidae, 215–216
    Prismatolaimus, 216–217, 216f–217f
Psilenchidae, 54, 55f, 65

Q

Qudsianematidae, 66

R

Reproductive system
  female
    columella, 58
    ovary, 57–58
    oviduct, 58
    spermatheca, 58
    types, 57, 58f
    uterus, 59
    vagina, 59
    vulva, 59
  male, 59, 60f
    primary sexual characteristics, 59–60
    secondary sexual characteristics, 60–61
  tail, 61–63, 62f
Rhabditida, 68–69
  Cephalobina, 69
    Cephaloboidea. *See* Cephaloboidea
    Panagrolaimoidea. *See* Panagrolaimoidea
Rhabditidae, 54, 55f
Rhabditid nematodes, 53–54, 54f
Rhabdolaimidae, 54, 55f, 64

S

Sand dune ecosystems, 232

T

Triplonchida, 211–212
  Diphtherophorina, 212

Diphtherophoroidea. *See*
    Diphtherophoroidea
Tobrilina, 213–214
  Prismatolaimoidea. *See*
    Prismatolaimoidea
Tripylidae, 64
Tripyloidea, 88–89
  Tripylidae, 89–90
    Tripyla, 90–92, 90f–91f
Tylenchida, 97–99
  Criconematina, 140–141
    Criconematoidea. *See* Criconematoidea
    Tylenchuloidea. *See* Tylenchuloidea
  Tylenchina, 99–100
    Anguinoidea. *See* Anguinoidea
    Dolichodoroidea. *See* Dolichodoroidea
    Hoplolaimoidea. *See* Hoplolaimoidea
    Tylenchoidea. *See* Tylenchoidea
Tylenchidae, 54, 55f, 65
Tylenchid nematodes, 51–52, 52f
Tylenchoidea, 101
  Tylenchidae, 101–103
    Basiria, 108–109, 109f–110f
    Boleodorus, 109–111, 111f–112f
    Filenchus, 104–106, 105f–106f
    Malenchus, 106–107, 107f–108f
    Tylenchus, 103–104, 103f–104f
Tylencholaimidae, 54, 55f, 67
Tylencholaimoidea, 194–196
  Leptonchidae, 197–199
    Leptonchus, 201–203, 202f
    Tyleptus, 199–200, 200f–201f
  Mydonomidae, 203
    Dorylaimoides, 203–205, 204f
  Tylencholaimidae, 196–197
    Tylencholaimus, 197, 198f
Tylenchuloidea, 144–145
  Paratylenchidae, 145
    Paratylenchus, 145–147, 146f
Tylopharyngidae, 54, 55f

W

Wetland ecosystems, 232

X

Xiphinematidae, 67

Printed in the United States
By Bookmasters